*Home Surroundings
the Waldorf Way*

Home Surroundings the Waldorf Way

A Guide for 1st and 2nd Grade Teachers

Roberto Trostli

Printed with support from the Waldorf Curriculum Fund

Published by:
Waldorf Publications at the
Research Institute for Waldorf Education
351 Fairview Avenue, Suite 625
Hudson, NY 12534

Title: *Home Surroundings the Waldorf Way*
  A Guide for 1st and 2nd Grade Teachers
Author: Roberto Trostli
Layout: Ann Erwin
Proofreading: Ruth Riegel

ISBN #978-1-943582-67-9

Copyright © 2022 by Roberto Trostli and Waldorf Publications

All rights reserved. No part of this book may be reproduced in any form or medium without written permission of the author. Only brief passages may be quoted for purposes of reviewing the book without written permission from the author.

Table of Contents

Introduction		7
I.	Home Surroundings in the Waldorf School	9
II.	Stories in First and Second Grades	17
III.	Fables & Animal Tales	19
IV.	Nature Stories	27
V.	Touching the Children's Souls: Stories with a Moral Element, Parables, Pedagogical Stories, and Birthday Verses	45
VI.	Composing Descriptive Nature Stories	58
VII.	Composing Stories with a Moral Element	66
VIII.	Composing Pedagogical Stories	69
IX.	Working with Birthday Verses	75
X.	Practical Aspects of Working with Home Surroundings	81
Appendix 1	Rudolf Steiner's Home Surroundings Curriculum Indications in the Original German	87
Appendix 2	"Nature Spirits" by Frits H. Julius	88
Appendix 3	A Selection of Birthday Verses	97
Appendix 4	Suggested Resources	113

Introduction

Of all the subjects that I taught in first and second grade, Home Surroundings posed the greatest challenge. As a student of Rudolf Steiner's educational work, I was surprised and dismayed that he gave only brief indications of what to cover in Home Surroundings and that he only spoke on a few occasions about how to approach nature study in the early grades. I was especially chagrined that Rudolf Steiner offered little guidance about how to work with "parables of nature." As a teacher of the upper grades, I had pondered these parables as they manifested in the physical sciences, but I didn't know how to recognize them in the world of the natural sciences.

In my conversations with other teachers, I found that some were entirely unfamiliar with Home Surroundings in the early grades, and that few of those who knew about it devoted two or three blocks to this subject. Some teachers did include nature stories in the early grades, though typically not in blocks, but no one I spoke with had a unified or comprehensive view of what we should teach or how to do so. In addition, I found only a few teachers who worked with pedagogical stories or birthday/report verses.

Faced with the challenge of developing my own curriculum and approach, I committed myself to
1) devoting two blocks per year in first and second grade to Home Surroundings,
2) researching and gathering nature stories of many types,
3) working with fables and animal stories in the way that Rudolf Steiner indicated,
4) composing as many of my own nature stories and poems as I could, and
5) deepening my ability to recognize parables of nature and bringing those to the students through pedagogical stories and birthday verses.

This book is the result of my work with the two classes of children that I taught from first through eighth grade. It contains examples of my original compositions to show teachers my approach and to reassure teachers that they too can compose original stories and verses.

I hope that this book will solidify Home Surroundings as a core subject in the early grades. I hope that teachers will be excited to explore this subject and will share their enthusiasm with others. And I hope that their Home Surroundings lessons will provide their students with a foundation for lives filled with meaning and purpose.

– Roberto Trostli
Richmond, VA, Summer 2021

I.
Home Surroundings in the Waldorf School

Home Surroundings is one of the three pillars of the main lesson curriculum of the first three grades. (The other two are Language Arts and Mathematics.) In spite of its importance, it remains one of the least understood and most underdeveloped subjects of the Waldorf curriculum. Unlike the other main lesson subjects in these grades, there is no consensus about its purpose, curriculum, or method.

Rudolf Steiner called this subject *Heimatkunde* or "knowledge of the homeland." Heimatkunde was an essential part of the elementary school curriculum of all German, Austrian, and Swiss schools in the 19th and early 20th centuries. It included topics in natural science, local history, and local geography and served as a preparation for these subjects in later grades.

English-speaking Waldorf teachers often refer to the subject as Nature Study. I consider this name too narrow because Home Surroundings also includes local geography, history, occupations, and other related subjects. I have therefore decided to call it Home Surroundings partly in tribute to Rudolf Steiner, but mostly because that name fully expresses the intention of this subject: to help our students feel at home in the world.

Home Surroundings in the early grades strives to orient students to their surroundings; it strengthens and deepens their connection to the natural world and to the world that human beings have created. Like its original European counterpart, Home Surroundings prepares students for further explorations in natural science, geography, history, and the physical sciences. What distinguishes Home Surroundings in the first two grades from what is studied later is that the subject is built on an imaginative foundation rather than on scientific observation or causal, analytical thinking.

To understand the role of Home Surroundings in the early grades, it is useful to examine how children's relationship to the world changes during the course of their education. This changing relationship is beautifully summarized by William Wordsworth in his "Ode" on Intimations of Immortality, from *Recollections of Early Childhood*, a portion of which is reprinted here.

There was a time when meadow, grove, and stream,
    The earth, and every common sight,
        To me did seem
        Apparelled in celestial light,
The glory and the freshness of a dream.
It is not now as it hath been of yore;—
    Turn wheresoe'er I may,
        By night or day.
The things which I have seen I now can see no more...

Our birth is but a sleep and a forgetting:
The Soul that rises with us, our life's Star,
    Hath had elsewhere its setting,
        And cometh from afar:
    Not in entire forgetfulness,
        And not in utter nakedness,
But trailing clouds of glory do we come
    From God, who is our home:
Heaven lies about us in our infancy!
Shades of the prison-house begin to close
    Upon the growing Boy,
But he beholds the light, and whence it flows,
    He sees it in his joy;
The Youth, who daily farther from the east
    Must travel, still is Nature's Priest,
        And by the vision splendid
        Is on his way attended;
At length the Man perceives it die away,
And fade into the light of common day.

Early Childhood: Initially the earth is an unfamiliar place for young children. In the preschool, children are helped to feel increasingly "at home" in the world through festivals, verses, songs, movement exercises, stories, and play. None of these brings children into a conscious relationship with their environment. Rather, children simply participate in activities that link them to their surroundings and experience the world primarily through their imagination and their will. By not bringing their experiences to consciousness, we allow children to continue to see the world with "eyes of spirit."

Grade School: Children's spiritualized perception of the natural world begins to wane once they enter the grade school. The Home Surroundings curriculum can extend some of this perception during the early grades, but as the children mature, they increasingly experience the world with their feelings, with "eyes of soul." The stories and descriptions of nature in the middle grades stimulate children's curiosity and strengthen their interest in and sense of connection to the world.

Middle School and High School: As they mature and begin to think more objectively, children feel increasingly divorced from their environment and eventually view it mostly with physical eyes ("eyes of the body"). An objective view of the world is achieved through their thinking, and this view gains meaning as it is infused with understanding.

From this overview we can see that the Home Surroundings curriculum in the early grades has a special task: to link children's spiritualized experience of nature with their soul perception of the natural world. This soul-spiritual relationship to the world will serve as a foundation for students' future geographical and scientific studies and for a lifelong meaningful connection to the world.

Home Surroundings in the Waldorf Curriculum

Rudolf Steiner spoke about the Home Surroundings curriculum in the First Lecture on the Curriculum which is included in *Discussions with Teachers* (1919). Here are his indications*:

> **First Grade:** To what has just been described [telling and retelling of fairy tales, legends, and accounts of outer realities] must be added what can stimulate children to reflection; you explain to them what lies near at hand and this will later be brought to them again in an orderly way as Geography and Natural History. These subjects are brought close to their understanding by linking them to things familiar to them—plants, animals, configurations of the land, mountains, and rivers. This is called in school "Study of Home Surroundings," knowledge of the homeland. But the point is that we bring about a certain soul-awakening in the children just in this very first year of their school life, an awakening as regards their environment, so that they learn to connect themselves with it.
>
> **Second Grade:** At the beginning of the second grade, we will continue with the telling and retelling of stories and try to develop this further. Then the children can be brought gradually to the point of writing down the stories we tell them. After they have had some practice in writing

---

* For the first grade section, I have chosen to use the translation included in Stockmeyer's Curriculum because it is closer to the original German. The second and third grade sections are from *Discussions with Teachers*.

down what they hear, we can also have them write short descriptions of what we've told them about the animals, plants, meadows, and woods in the surroundings.

Third Grade: You have seen how we make free use of familiar things from the immediate surroundings for our independent instruction in general knowledge. In the third grade, when the children are going on nine, it is quite possible for this instruction to provide them with an idea of how mortar is mixed, for instance—I can choose only a few examples—and how it is used in building houses. They can also have an idea of how manuring and tilling are done, and of what rye and wheat look like. To put it briefly, in a very free way, we allow the children to delve into the elements of their immediate surroundings that they are capable of understanding.

We can see that the path of the curriculum mirrors the child's path of incarnation from cosmic beginnings down to earth. The topics in the first two grades orient children to their surroundings, while the third grade curriculum helps the child make sense of human activities and occupations. What we must not forget is that Home Surroundings is intended to bring about a "soul awakening" in the children. It also has the potential to deepen and strengthen the soul bonds between teacher and students.

Introducing Children to Nature

In *The Kingdom of Childhood* (1924) Rudolf Steiner spoke extensively about how to teach young children about the natural world. The following excerpts provide us with invaluable guidance.

> Our teaching and education are to be built, then, on imagination. You must be quite clear that before the ninth or tenth year the child does not know how to differentiate itself as an ego from its surroundings. Out of a certain instinct children have long been accustomed to speak of themselves as "I," but in truth they really feel themselves within the whole world.... Not until the age of nine or ten do children really learn to distinguish themselves from their environment. This is something you must take into consideration in the strictest sense to give your teaching a proper basis.
>
> Therefore it is important to speak of everything that is around the children—plants, animals, and even stones—in a way that all these things

talk to each other, that they act among themselves like human beings, that they tell each other things, that they love and hate each other. You must learn to use anthropomorphism in the most inventive ways and speak of plants and animals as though they were human. You must not "ensoul" them out of a kind of theory but treat them simply in a way that children can understand before they are able to distinguish between the lifeless and the living. …Indeed, you can treat all things that feel and live as if they were people, thinking, feeling, and speaking to one another, as if they were people with sympathy and antipathy for each other. Therefore everything that you bring to a child of this age must be given in the form of fairy tales, legends, and stories in which everything is endowed with feeling. It must be kept in mind that nourishing the instinctive soul qualities of imagination in this way is the best foundation for the child's soul life.

For this task, as a teacher, you must have artistic feeling in your soul and an artistic disposition. It is not only what you think out or what you can convey in ideas that works from teacher to child, but, if I may express myself so, it is the imponderable quality in life. A great deal passes over from teacher to child unconsciously. The teacher must be aware of this, above all when telling fairy tales, stories, or legends full of feeling. It can often be noticed in our materialistic times how a teacher does not really believe what he or she is telling and looks on it as something childish. It is here that Anthroposophy can be the guide and leader of a true knowledge of the human being. We become aware through Anthroposophy that we can express a thing infinitely more fully and more richly if we clothe it in pictures than if we put it into abstract ideas. A child who is healthy naturally feels the need to express everything in pictures and also to receive everything in picture form. (30-32)

It will always be a question of finding out what child development demands at each age of life. For this you need real observation and knowledge of the human being. Think once again of the two things that I have discussed, and you will see that children up to their ninth or tenth year are really demanding that the whole world of external nature be made alive, because children do not yet see themselves as separate from this external nature; therefore we tell them fairy tales, myths, and legends. We invent something ourselves for the things that are in our immediate environment in order that in the form of stories, descriptions, and

> pictorial representations of all kinds we may give children in an artistic form what is found within their own soul, in the hidden depths that children bring with them into the world. And then after the ninth or tenth year, let us say between the tenth and twelfth year, we introduce children to the animal and plant world as has been described. (50)

In Lecture 4 of *The Kingdom of Childhood* (1924), Rudolf Steiner introduces a story by speaking about the benefits of teachers staying with students for as many years as possible so that they can refer to what they have taught previously and consider it on a different level.

> Let me therefore give you an example of something that can sink into the child's soul so that it grows as the child grows, something that you can come back to in later years and use to arouse certain feelings. Nothing is more useful and fruitful in teaching than to give the children something in picture form between the seventh and eighth years and later, perhaps in the fourteenth and fifteenth years, to come back to it again in some way or other. Just for this reason we try to let the children in the Waldorf school remain as long as possible with one teacher. When they come to school at seven years of age, the children are given over to a teacher who then takes the class as far as possible. For it is good that things that at one time were given to the children in germ can again and again furnish the content of the methods used in their education. (57)

Rudolf Steiner then tells a story about a violet and the sky. He makes extensive comments about how to work with the story, and I recommend that teachers read the entire section of the lecture.

In *A Modern Art of Education* (1923) Rudolf Steiner spoke about the importance of having children experience something fully before developing ideas or thoughts about it.

> The whole process of development is ruined if children are led immediately into abstraction—if we teach them to do something by giving them purely mental concepts. Healthy growth, on the other hand, is always the result of introducing an activity first, followed by the idea, which develops from the activity. Reading is essentially mental, and if it is taught before writing, children prematurely develop only the head instead of the forces of their whole being.
>
> Methods such as these can guide education into an area that embraces the whole human being: the realm of art. This must, in fact, be the goal

of all teaching up to the age of about nine and a half. Image, rhythm, measure—these qualities must imbue all our teaching. Everything else is premature.

Consequently, before this time it was impossible to convey anything to children that strongly distinguishes between them and the outer world. It is not until around the ninth or tenth year that children begin to realize they are beings apart from the outer world. Hence, when they first come to school, we must make all outer things seem alive. We speak about the plants as living beings, conversing with us and one another in such a way that the children's view of nature and people is full of imagination. Plants, trees, and clouds all speak to children, and at this young age they must not feel separate from the outer world. In an artistic way, we must give children the feeling that just as they can speak, everything that surrounds them also speaks.

During these early years, the more we respond to the needs of children's innermost being, the more we enable them to flow into their environment. We vividly describe plants, animals, and stones so that articulate spirituality wafts toward them in weaving imaginations. These are the years when the feeling of children's souls must flow into their breathing, blood circulation, and into the whole vascular system—indeed, into the whole human organism. If we teach in this way, children's feelings will be called on in a way that is appropriate for today. Thus, children develop naturally in body and soul. (124-125)

Having stressed the importance of imaginative, pictorial instruction, Rudolf Steiner warns of the damage that is caused by premature intellectualism.

In *The Renewal of Education* (1920) he states:

Everything you teach in the way of scientific descriptions is damaging before the age of nine. Realistic descriptions of beetles or elephants or whatever, in the way we are used to giving them in the natural sciences, are damaging children before this age. We should not work toward a realistic contemplation, but toward imagination. (95-96)

In *Soul Economy and Waldorf Education* (1921), Rudolf Steiner addresses the issue even more dramatically:

After the change of teeth, when children experience conceptual thinking, it is as if spikes were being driven through their whole being, especially when such concepts come from the inorganic, lifeless realm. Anything taken from the soulless realm will in itself estrange a child. Consequently, those whose task is to teach children of this age need an artistic ability that will imbue everything they bring with life; everything must be alive. Teachers must let plants speak, and they must let animals act as moral beings. Teachers must be able to turn the whole world into fairy tales, fables, and legends…. (161)

Keeping this orientation in mind, let us now consider the types of stories that we should include in our Home Surroundings blocks in the first and second grades.

II.
Stories in First and Second Grades

Because children in the early grades are so imaginative, Rudolf Steiner encouraged teachers to use stories as a primary means of instructions. These stories include fairy tales, legends, fables, animal tales, nature stories, pedagogical stories, and parables.

First graders hear stories with a dreamy consciousness, so the stories they hear should have archetypal themes. By second grade, children are beginning to listen with a more wakeful consciousness, so the stories they hear can have more realistic themes.

Rudolf Steiner stressed the importance of fairy tales in first grade when he said, "We avoid using passages that do not stimulate the imagination and make as much use as possible of texts that activate the imagination strongly, namely, fairy tales—as many fairy tales as possible." (*Practical Advice to Teachers*, 169)

Why should we tell fairy tales? According to Rudolf Steiner, fairy tales, like myths, represent cosmic realities in earthly form. In ancient times, people could still perceive these realities through their innate clairvoyance. Fairy tales contained images that were relics of this clairvoyance, and even though they might not be completely understood, those who hear them recognize that they convey deep truths. Because fairy tales express spiritual wisdom in pictorial form, they are very appropriate for young children.

Fairy tales express what it means to be a human being in the various roles we assume in our lives. They portray archetypal human experiences: what it means to be a mother or father, daughter or son; what it means to come of age, to seek our destiny, to be tested and to persevere until we achieve our goal.

Fairy tales offer us a unique opportunity to bring meaning and healing to our students' lives. They engender a sense of satisfaction in the children because good is rewarded and evil punished. As G.K. Chesterton put it, "For children are innocent and love justice, while most of us are wicked and naturally prefer mercy."

In terms of sources for fairy tales, legends, fables, animal stories, and nature stories, we have to consider whether we want children to hear the stories more unconsciously or more consciously. I think that it benefits first graders to dream their way into the fairy tales because that way the archetypes can slowly emerge from their unconscious into their conscious minds. Fables and nature stories, on the other hand, should be received with a more wakeful consciousness because we want children to relate to them personally.

For this reason, I recommend using fairy tales from a single culture for the entire first-grade year. Because stories from a single culture have a consistent language and imagery, children will listen to them more unconsciously than stories from a variety of cultures whose unfamiliar aspects tend to wake children up. In contrast, for legends, fables, animal tales and nature stories, I think that teachers should use many different sources. This will stimulate the children's interest and engage them more fully.

According to Rudolf Steiner, the best source of stories for children are their teachers who know them and know what they need to hear. I strongly encourage teachers to write their own nature stories, pedagogical stories, and parables because these will touch their students most deeply. For many teachers, composing such stories is uncharted territory, but those of you who embark on this journey will find that your efforts are richly rewarded.

A footnote note about including elemental beings in nature stories: Rudolf Steiner made many statements about elemental beings in his lectures which help us recognize the reality and importance of these beings. Teachers who casually refer to gnomes or to fairies of various kinds do these beings a disservice because elemental beings perform important tasks for human beings and for the world.

I recommend not naming or describing elemental beings in nature stories, but when necessary, referring to them by more general terms such as "Brother Wind's helpers." Children who still experience these beings should not be affronted by our references, and those who have moved beyond their childhood clairvoyance will not be put off by sentimental terminology.

Educating oneself about elemental beings adds depth and meaning to our work in Home Surroundings, especially for what we present in first grade. I recommend Marjorie Spock's *Fairy Worlds and Workers* and working with Frits Julius's article "Nature Spirits" (reprinted in Appendix 2). Both of these works help us form a deeper connection to the hidden worlds of nature.

III.
Fables and Animal Tales

Fairy tales portray archetypal situations of our journey through life. A child may identify with a fairy-tale character, but the identification remains largely unconscious. In contrast, the stories told in second grade help children get to know themselves and other people. One way to achieve this goal is to tell legends of people who have realized their noblest qualities. Such legends are an important part of the second grade curriculum, but since they are not included during Home Surroundings blocks, we will not discuss them here.

In contrast to legends of virtuous people, fables and animal tales reveal aspects of human beings' lower nature. Although these stories use characters drawn from nature, their intention is to depict human faults and foibles. Fables and animal tales should help students recognize positive and negative human traits and provide guidance—often through negative outcomes—for appropriate behavior.

There are myriad sources for fables and animal tales from many cultures, and I do not recommend any particular ones. I do, however, think that students benefit from hearing groups of stories from a common cultural source or groups of stories from different cultures that portray the same animal or analogous situations. Teachers should try to find the stories that will speak to the class and engage their sense of humor. With some searching, you should be able to find stories that address all the current needs of the class and of particular students.

Fables and animal stories do not have to be presented only in Home Surroundings blocks; they can be presented as literature in Language Arts blocks or as stories told at the end of Math main lessons. Fables and animal stories can also function as pedagogical stories and be used throughout the year to address the students' needs.

The clearest directions for working with fables the Waldorf way are given in *Discussions with Teachers* (1919). Rudolf Steiner gives detailed instructions about how to present and work with fables, and he presents several examples. He rejects the common approach to analyzing a story after reading or hearing it. Instead, he tells teachers to speak about the fable *before* telling it so as to appeal more to the children's feelings rather than their understanding. Discussing the fable with the children first, he says, will give them a sense of completion and contentment when they hear it told.

In Lecture 6, Rudolf Steiner demonstrates how he would speak with the children in preparation for telling them a fable. He vividly describes the animals in the fable and refers to what the children might already know about the animals. Then he draws a connection between the animals' characteristics and behavior and human character and behavior. Once the children have been introduced to the characters and themes, they are ready to hear the story.

Excerpt from Lecture 6 of *Discussions with Teachers* (1919) with Commentary

Rudolf Steiner begins by showing how he would speak with the children in preparation for telling them a fable. Note how directly he speaks to the children, how he uses humor, and how skillfully he calls on the children's memories and perceptions to help them build vivid imaginative pictures.

> I would therefore take something like the following with the children (but you would have to work it out in greater detail). I would say: "Now look, boys and girls, you have certainly seen some dogs at some time or other in your life. If any of you have never seen a dog, then you must have been hiding in some dark corner! And you must have noticed that not all dogs are alike. They are very, very different from one another. There are tiny little dogs, small dogs, larger dogs, and great, big dogs. You have probably always been afraid of the very large dogs; but you have not been afraid of the tiny little dogs—or maybe you have, because sometimes they bite people's calves.

> Now today we will look at some of these dogs. You have probably often seen a meat cart in the streets pulled by a butcher's dog. If you have looked carefully, you have probably noticed that the rest of the time this dog sits in front of the butcher's stall and makes sure no one steals the meat. If anyone comes who isn't allowed and takes the meat, the dog must bite that person, or at least bark. Now, you see a butcher's dog cannot be a tiny little animal; no, he must be a big dog. You will also notice that small dogs are never harnessed to a butcher's cart, nor do they watch a butcher's stall.

Next Steiner makes the connection between animals and human beings. This is a key element of this conversation. Note how he explains a familiar colloquial expression as another way of helping them connect with the story.

> You can compare a butcher's dog with a person who has to guard
> something. You can often compare animals with people. Animals have
> to do things through instinct, and people must often do the same things
> because it is their duty. People and animals have to do similar things and
> therefore they can be compared. Suppose a man has to guard something
> just like a butcher's dog does at the meat stall; the man will form a certain
> habit. If someone comes and tries to steal something, he will take hold of
> him by the hair. Yes, when someone is doing something wrong—you take
> hold of such a person "by the comb." But a person has hair, not a real
> comb. You pull the hair, and that hurts, so the person doesn't try to get
> away; that is why you do it. You don't say this kind of thing point blank.
> If you said straight out, "I will pull your hair," it wouldn't be as much fun.
> There must always be a little fun in life, so you say that you take someone
> "by the comb." A person has hair, and is sometimes insolent; a rooster
> is almost always insolent, and has a comb; that's why you say, "I'll take
> you by the comb." You can imagine that if, for example, another insolent
> creature came along, wanting to take a piece of meat out of the stall, the
> butcher's dog might say, "I'll take you by the comb!" Then you would have
> made a very good comparison between a person and a dog.

Rudolf Steiner proceeds to make judgments about the animals in the stories; this will help the children interpret the story properly when they hear it. In this section he mentions a proverb that sheds further light on the meaning of the story. Steiner concludes this section with an important comparison between animals and people.

> Now you know, children, there are also other dogs, small dogs, who are
> mostly lazy creatures; they are miserably lazy. They lie on cushions or
> sometimes even on laps. Basically, they are idle fellows. They are "cushion-
> dogs," those lapdogs.
>
> They are not as useful as the butcher's dog. The butcher's dog is of some
> service; the lapdogs, they only play; they are basically useless. But if
> anyone does anything wrong, the butcher's dog will take that person by
> the comb—the dog will seize that person and give a thorough shaking.
> This is of some use, because the other creature will not be able to steal
> the meat. The lapdog doesn't do anything useful like that, but only yaps,
> yaps at everybody; and especially when big dogs come along, the little
> lapdog rushes out and yaps and yaps and yaps. But their bark is worse
> than their bite; that is what the proverb says and that is what the large
> dogs think as they pass by. You can also see how large dogs go by very

calmly; they let the little yappers yap, and think to themselves: Yapping dogs don't bite, they are not brave, they are cowardly. But a butcher's dog must always have courage. The lapdogs run after the big dog and yap, but if the butcher's dog turns round and looks at them, they immediately run away. So you see these little dogs are certainly lazy; they only do what is unnecessary and they are good for nothing. They are like certain people whom we should not listen to, even though they very often yap at us.

Steiner continues to describe the dogs in terms of their size and the work they do. He ends the section with humor.

These lapdogs are very small, the butcher's dog is large. But there are other dogs in between—not as big as the butcher's dog, but larger than the lapdog. Among the medium-sized dogs is the sheepdog. The sheepdog has to guard the sheep. In many districts it is a more difficult job than it is here. In many places—in Russia, for example—there are wolves, and the dog has to keep the wolves or any other animals away from the sheep; and so the sheepdog has gotten into the habit of continually running round the flock. In our country too it is good to have dogs who run round the flock all the time, because the shepherd is often asleep, and any evil-minded creature could come and steal some of the flock. So the sheepdog runs round and guards the flock. Even when there are no wolves, it is good for the sheepdog to run round and guard the flock; and sometimes the sheepdog guards the shepherd, as well, who is then awakened. It might even happen that a shepherd could be stolen while asleep!

In the last section, Steiner makes the comparison between the dogs and human beings ever more explicit so that the children will understand the meaning of the passage when they hear it without any further explanation.

Thus the shepherd's dog, the sheepdog, is of service; the sheepdog is a useful animal and can be compared with people who have found their proper work in life, people who are not useless like the idlers, the lapdogs. Yes, in human life too there is this difference between those who are like sheepdogs and those who are like butcher's dogs. They are both useful, although the latter, like the butcher's dog, are at times rather rough. Sometimes they say exactly the right thing in a few words and straight to the point, to guard something, to ward off an enemy. You can make a comparison with the sheepdogs also; they are like people who work quietly, waiting calmly until difficulties are upon them. The sheepdog runs round and round for a long time; he has nothing to do, but he must always be prepared for action, so that when the wolf or another enemy appears, the sheepdog will be strong and courageous and well prepared to attack at the right moment. There are also people who have the duty to watch and wait until they are called on to fulfill some task. They must not allow themselves to be harassed by petty things in life, but always have to be ready for the moment when they must act, and act correctly. This is how I would speak to children, choosing some particular example from the animal world and leading their thoughts to analogies between animals and people. After speaking somewhat in this way, you can read aloud the following passage, and when you have read it, explanations will be unnecessary. If you were to give the children this little story first without any explanation they would not be fully prepared, because their perceptions and feelings would not be directed to what it contains. If on the other hand you do not explain until afterward, you would pedantically pull the passage apart, and so they would not be able to read it properly either.

Having completed the introduction, Steiner then tells the fable.

The Sheepdog

One evening an old sheepdog who was a faithful keeper of the master's sheep was on the way home. As he went down the street, some little lapdogs yapped at him. He trotted on without looking around. When he came to the meat stall, a butcher's dog asked him how he could stand such constant barking and why he didn't take one of them by the scruff of the neck. "I won't," said the herdsman's dog. "None of them is worrying me or biting me. I must save my teeth for the wolves."

Steiner concludes by saying, "You do not need to say another word to the children. The preparation must come first so that they understand what is told." (*Discussions with Teachers*, 71–75)

Rudolf Steiner's indications may be summarized as follows:
1. Have a conversation with the children about the story's characters and themes before reading a passage or telling a story.
2. Speak to the children directly in a straightforward manner.
3. Call on the children's memories and perceptions about the animals.

4. Describe the animals and their actions vividly.
5. Compare and contrast the animals in the story with human beings.
6. Share judgments about the nature and actions of the animals and human beings.
7. Elucidate colloquial expressions and proverbs.
8. Use humor.

All of these techniques help us touch the children's souls so that they will have a joyful, meaningful, and satisfying experience of the story.

Later in Discussion 6, Rudolf Steiner demonstrates how to introduce a nature poem, and in Lecture 4 of *The Kingdom of Childhood* (1924), he shows how to work with a nature story and includes suggestions for review, working with the temperaments, and more. I recommend that teachers look up these excerpts because there is much to be learned from Rudolf Steiner's example, suggestions, and explanations.

Telling Fables

Likewise I encourage teachers to follow Rudolf Steiner's example when telling fables or animal stories. In order to do so, you need to inform yourself fully about the animals in the story and provide vivid descriptions so that children can get to know the animals and relate to them.

You also need ponder the animals and situations in terms of human beings. If we can help the children recognize the connection between animal and human characteristics and behavior, they will come to a better understanding of themselves and a deeper appreciation for others.

With Rudolf Steiner's indications in mind, here is how I presented the fable of "The Fox and the Crow."

Today I am going to tell you a story about an animal that several of the students in our class like best: the fox. As you know, foxes are related to dogs: foxes are shaped much like a thin dog. They have narrow faces, pointed ears, and their pride and joy: their beautiful bushy tail. Foxes are hunters, who rely on their senses to hunt, especially their hearing. Their hearing is so keen that they can hear a mouse squeak from a hundred feet away. When a fox hears a mouse, she approaches it very slowly and silently and then suddenly pounces on the mouse, trapping it with her paws.

Foxes typically have six to ten pups or kits. For the first few weeks, the kits drink milk. Then they eat food that their parents catch and chew up for them. Once they begin to eat meat, they are ready to learn to hunt. The mother or father fox will catch a mouse or other small creature and bring it back to the den alive so that the kits can practice pouncing on it.

Some people like to hunt foxes because they are such cunning prey. When a fox is being chased by dogs, she uses many tricks to elude them. She might walk backward over her own tracks and then leap to one side so that when the dogs follow her trail, they come to the end and don't know which way she went. At other times a fox might walk along a streambed so that the dogs lose her scent. Foxes have been known to jump onto fences and walk along the top rail so that the dogs cannot follow. Rarely does a fox allow herself to be outwitted.

Crows are also very clever animals, among the cleverest of all the birds. Unlike the fox who lives only with her immediate family, crows live in large flocks that nest high in trees. Unlike the fox who can slink silently past us, we always know when crows are around because of their loud caws. Crows are proud birds who like to be noticed. Whenever a crow has something to crow about, the whole flock hears about it and responds with raucous cries.

Crows are constantly on the lookout for food. They will eat almost anything—seeds, insects, other birds' eggs, and even little animals. They don't like to work any harder than they have to for their meal and are happy to steal any food that they can find. So if you leave some food on your picnic blanket and go off to play, you might find that some crows have come along and flown away with your feast.

Crows build large nests out of branches and twigs, and they like to collect objects that they keep in their nests. They especially like shiny objects, and a crow's nest will be filled with little treasures.

Now, we all know people who are clever like foxes. They use their wits rather than their strength, and when they want something, they think about it and figure out the best way to accomplish it. Such people enjoy being alone because it gives them time to think about what they want without too many distractions.

We all also know people who are a bit like crows—they like to be with others and care about what others think of them. They can be loud and boisterous and they want to be noticed. Such people like to have and even to hoard things. They will do whatever they have do to get what they want. Such people tend not to think too much before they act, and they can sometimes be foolish.

Here is an old fable about a fox and a crow:

The Fox and the Crow

One day a fox was walking along the woods when he spied a crow sitting on a branch in a tree. The crow was holding a large piece of cheese in his beak and was about to eat it. "O Crow," the fox said, "what a marvelous bird you are! You have such strong wings and such beautiful shining feathers. You are so brave and strong, and you fly like the wind. Why, you are so magnificent, you should be the King of the Birds!"

The crow looked down at the fox but didn't say anything. "But what about your voice?" said the fox. "Is it the voice of a king? Is it as strong and as proud and as beautiful as you are?"

The crow opened his beak and gave a loud caw. As he did, the cheese dropped and fell into the fox's waiting mouth. The fox ate the cheese and said, "You do have a loud voice, and perhaps you would make a fine king, but what you lack is wit."

We would review and discuss the story the next day. During the review, the students would retell and perhaps act out the story, and then we would discuss it. I would expect that some children would not be shy about identifying themselves as more foxlike and pointing out some of their classmates' crow-like qualities. I doubt that many would admit to identifying with the crow, but they might inwardly recognize that they share some of the crow's characteristics.

Although these kinds of conversations before sharing a story may feel awkward to teachers who are accustomed to telling stories without a preamble, I encourage you to practice this technique (and also utilize it to introduce other works of poetry and prose), for it will yield unexpected rewards.

IV.
Nature Stories

Descriptive nature stories depict the origins or the characteristics of the heavens, weather, the seasons, the landscape, bodies of water, plants, and animals. One category of these stories is called "pourquoi" stories (*pour quoi* means "why" in French). They describe what the world and its creatures are like and how they came to be that way. Pourquoi stories are often fanciful and light-hearted, giving a humorous explanation for a name, appearance, characteristic, or behavior.

Another category of descriptive stories is narrations of the world and its creatures. These stories are the most realistic type of nature stories, but we do not tell them simply to convey information; we use them to stimulate the students' interest in and connection to their characters. Descriptive stories can be told as simple narratives, but framing them as biographies, recollections, or conversations makes them more interesting and engaging.

I think that first graders are best served by stories that are more cosmic in scope—stories about the heavens and the heavenly bodies; the seasons and the weather; the earth and its landscapes; and the ocean and other bodies of water. Because these aspects of the natural world are the grandest, stories with these themes allow the children to orient themselves to and connect with the world in the widest and most general sense.

Because second graders are more down to earth than first graders, they benefit from descriptive stories which have specific themes. These include origin stories, stories that account for natural processes or cycles, and stories about particular aspects of the landscape or about particular plants or animals. Second graders will appreciate the details of stories that make the world and its creatures more familiar and comprehensible. The descriptive stories told in the first two grades lay an imaginative foundation for more objective nature studies and for future work in geography, earth science, and meteorology.

Teachers will be able to find pourquoi and descriptive stories from many different cultures, each with its distinctive vocabulary and imagery. When choosing among them, let your knowledge of your students guide you, especially your sense of which stories will speak to their sense of humor.

Rudolf Steiner encouraged teachers to compose their own nature stories because the children are much more receptive to, and more deeply affected by, what arises from their teacher's soul. I join him in encouraging you to attempt this task. You may well find that you are helped by the spiritual beings who are interested in you and your students and that the act of creation is blessed in unexpected ways.

Descriptive stories are the easiest to write and are probably the best type to try to compose at first. They require little more than dramatizing or personifying an aspect of nature. While it is more difficult to compose effective origin or natural cycle stories, I encourage you to try to your hand at these as well. Your descriptions and explanations do not have to be totally accurate in scientific terms; they just have to be true in human terms.

When working with descriptive stories, we can use the same pre-telling technique that we use for fables. Alternatively, the stories can be told more like fairy tales or legends, with a brief introduction. If at all possible, we should tell—rather than read—most of our stories, especially stories with a moral and pedagogical stories because children are much more deeply affected by what we say than by what we read.

I composed a handful of nature stories for first and second grade. Never having written children's stories, I approached the task with great trepidation. The first stories did not come easily, and I was self-conscious in the telling, but as the children responded joyously to my efforts, I became more confident. I have included a selection of my stories throughout this book to serve as examples for emulation or adaptation, or, if it suits a teacher's need, simply told as written.

First Grade Stories

Autumn

This story paints a picture of the Autumn season by describing different creatures' preparations for Winter in New England. I chose different creatures and environments for contrast and employed the refrain, "It was late Autumn, and the last days of warmth and sunshine had fled," to give the story structure and continuity.

This was the first story that I composed for our first Home Surroundings block, and I was still a bit uncomfortable with personifying the characters.

> It was late Autumn, and the last days of warmth and sunshine had fled. In her hive, the honey bee was making her final preparations for Winter. How busy she had been all Spring, seeking out fragrant flowers, gathering their pollen and nectar, and carrying them back to the hive for honey. All Summer long she had worked building the hive, making the countless six-sided rooms to store the honey for Winter food. As the days grew colder, she had helped narrow the entrances to the hive so that the North Wind could not blow in and freeze the bees. After one

last great meal of honey, the bee huddled with her sisters for warmth and slowly fell into a drowsy sleep. She would sleep through the short cold days ahead. If the Winter sun shone warmly and the snow thawed, perhaps she would awake. But looking out over the snowy land, she would see no reason to leave the warm hive. After another meal of honey, she would go back to her snug bed with her sisters and sleep again until the Spring sun would bring back the sweet flowers.

It was late Autumn, and the last days of warmth and sunshine had fled. Soon the frost would lie thickly on the grasses, and the edges of the pond would freeze. On the bottom, buried deep in the soft mud, lay the bullfrog. He had been born that Spring from a tiny translucent egg that had floated among the water weeds. He had become a shiny black tadpole with a pointy tail, wiggling through the waters, exploring his new home. Soon he had begun to change—sprouted back legs, then front legs, and his tail had fallen off! Now he could no longer live under water, but had to come to the surface to breathe. All Summer long he had basked in the sun by day, lazily floating half-submerged in the warm water. In the evenings he had joined the frog chorus, croaking and calling with full voice. Now he was tired and sluggish. He had made a small burrow in the mud and had settled in. Soon he would fall asleep, a sleep near unto death. On the bottom he would lie, barely breathing, barely living, until the warm sun would melt the ice on the pond, and life would start to stir again.

It was late Autumn, and the last days of warmth and sunshine had fled. With them had fled the barn swallow, for she could not stand the cold. In the Spring she had flown north in a great flock of swallows, until she had come to the familiar fields of the farm where she had been born. She and her mate had built their nest out of mud, and there they had warmed the eggs and raised their two little ones. All Summer long she had flitted and flown over the meadows and swamps, feeding on small flying insects. As the days grew shorter, the swallow had begun to feel restless and impatient. Other swallows had gathered, calling and chattering with each other about the journey ahead. Together they had flown southward to the lands where the snow never falls. There they would settle and wait until the Winter was past before returning north again.

It was late Autumn, and the last days of warmth and sunshine had fled. As the last few leaves fell to the ground, the mother bear had gathered them, readying a bed in her den for Winter. All Spring and Summer she had wandered through the meadows and woods, eating berries, fruits, and grubs. She had scooped fish and frogs out of the streams and had chased and caught chipmunks. One day she had found a hive of wild bees and had gorged herself with honey. She had grown round and plump, with a fat layer of blubber underneath her thick warm fur. As the days grew colder, she had begun to feel tired, less eager to wander and explore.

All she wanted to do now was sleep. When the snows came, she would walk outside for the last time, then settle down in her bed of rustling leaves, and fall into her deep Winter sleep. Many weeks would pass before the pangs of hunger and thirst would stir her awake. She would wander outside, foraging for food, the memories of Summer's plenty dim in her mind. Perhaps her footprints would be seen near the houses where the garbage cans of food beckoned. Then she would return to her den and her warm bed, and sleep again, waiting for Spring, for the time of year when life quickens and stirs.

Spring Is Coming!

By the second Home Surroundings block, I was more comfortable with composing my own stories and experimenting with different forms. Because the approaching season seemed so lyrical, I decided to write a prose-poem to celebrate the signs of Spring in New England. I was inspired to use the refrain "Spring is coming!" because it echoed a song by Elisabeth Lebret that we were singing in morning exercises. To enhance the story's lyricism, I used alliteration and assonance as much as I could. I used the present progressive tense to create a sense of immediacy and spoke directly to the children to awaken their sense.

Spring is coming! Spring is coming! What are the signs of Spring?

Listen to the drip, drip, drip of melting snow as it runs down the roofs in rivulets. Listen to the brook as it babbles, breaking out of its frozen fastness. The ice cracks and crazes, finally falling into the swollen stream.

The frozen earth is thawing; the ground feels spongy underfoot, and our feet sink into the squishy mud that sucks at our boots and doesn't want to let go. Soon the plants will poke through the cold, muddy crust, shove aside the debris of Winter, and cover the earth with a new mantle of green.

Spring is coming! Spring is coming! What are the signs of Spring?

The air is fresh, so tender and moist! Gone are the chilling winds that rushed around the eaves at night, that rattled down the chimney and whipped the branches against the side of the house. The Spring breeze blows gently, caressing the young buds and shoots, calling them from their Winter hiding places.

Spring is coming! Spring is coming! What are the signs of Spring?

The trees and bushes, the hedgerows, orchards, and meadows are full of birds. Their songs fill the early mornings, sometimes with sweet warbling of love, sometimes with loud, insistent cries of defiance that keep others at a distance. The evenings are graced by wild geese flying on sturdy wings, heading northward to newly thawed lakes and ponds, where they will build new nests and bring forth their young.

Spring is coming! Spring is coming! What are the signs of Spring?

The little creatures who slept through the Winter are waking. The moist earth is full of their tracks. At twilight you might catch a glimpse of them searching and scampering, venturing out of their homes and hiding places, looking at the world now born again. Soon the spotted fawns will walk beside the doe, the fox kits will tumble and wrestle by their dens, and the big black bear will teach her young to fish in the placid pools.

Spring is coming! Spring is coming! What are the signs of Spring?

My heart is lighter, my gaze is brighter, there is new strength and spring to my step. Winter is gone. The snow has fled. The earth awakes again.

A Drop of Water

This story chronicles the miraculous journey of a drop of water. Its descriptions provide an imaginative foundation for study of the water cycle in the upper elementary grades. I used the names Mother Ocean, Father Sun, and Brother Wind to reinforce the idea that the drop of water is a member of a cosmic family. To create a sense of completion (and endlessness), I ended the story by having the drop of water coming back home to where the journey began.

Some parts of the water drop's journey echo *Paddle to the Sea* by Holling C. Holling. (Whatever you encounter or read can serve as inspiration!) In retrospect, I think that this story contains more detail than is necessary for first grade, but I couldn't resist including as many pictures as possible.

> It was Autumn, and on Mother Ocean's cool bosom lay a drop of water. Above her, Mother Moon shone softly and the stars twinkled merrily. Below her, Mother Ocean's currents swirled gently, carrying the drop of water on her way. All was peaceful and quiet, but the drop of water was wide awake, thinking of far-off lands whose shores she had never seen. Would she ever see the wide, wide world?
>
> Slowly the night passed and the pink dawn began to glimmer in the east. Soon Father Sun's rays were shining brightly upon the ocean, warming the cool waters.

Now the morning breeze sprang up as Brother Wind blew briskly over the sea, stirring up waves. Over and again the drop of water was lifted up, then swept down again. A wave tossed her high in the air, and she sparkled and shone in the light. Up again she was flung, but this time, she did not fall. How light she felt, how shimmery! Up and up she rose, carried by Brother Wind's soft breath. With her were many other drops of water, sisters and brothers, friends and strangers, rising in a fine mist above Mother Ocean.

Below her the waves grew smaller and smaller, until the drop of water could hardly see Mother Ocean's ruffled surface. How long or far she flew, she never knew, but later she remembered seeing shores and valleys, plains and mountains. How high she was—and how cold! Feeling herself growing stiff, the drop of water shivered in fear. She was chilled through. To the left she shivered, and to the right. Up and down she shivered, forward and back, but nothing could stop her from freezing.

Looking around her, she noticed that all of her fellows were frozen too, each into a delicate six-sided form. Each of her friends was different, with lacy arms or bold plates, but all of them had six sides. She looked down at herself. She had become a snowflake too! As a mountaintop loomed near, the drop of water felt herself falling, gently drifting down with countless other snowflakes. Softly she landed on the mountain, glad to rest again. Covered and surrounded by her friends, the drop of water fell into a deep, contented sleep.

Days and weeks passed, and still the drop of water slept on the mountain. One day she awoke. She was no longer cold and stiff. Father Sun had warmed her through, and her lacy form had melted. Through the layers of snow she oozed, wondering what lay below. Soon she touched something; it was hard and rough. She was lying on the earth. Into a tiny hole she slipped, and she explored the small channels underground. How strange and dark they were—and how beautiful! All around were the glistening lights of quartz and mica. Here and there were tiny specks of gold dust. Everywhere she looked she saw delicate root hairs of plants, soaking up water for the plants to drink.

Down another slanted channel she slipped. Then all of a sudden the drop of water came out of the side of the mountain and ran along a little streambed. It was lined with tiny pebbles, worn by the Spring streams of bygone years. Down the mountainside she rushed, now laughing for joy, now gasping in fright, as she sped around sharp, sudden turns. For weeks she coursed down the mountainside, feeling the stream growing in strength beneath her. The stream roared round boulders, pushing them out of the way, sending them off with a crash. It tore at tree roots, digging away at the earth with strong and eager fingers. Nothing could resist its flow.

Finally the drop of water entered a river, swollen with Winter snows and Spring rains, but traveling more slowly than the stream. Here she had a chance to look at her surroundings: the tall trees, the fertile farm fields, and here and there the houses on whose porches children played. One night she passed under a bridge, and then she saw the dazzling lights of a city sparkling above her.

The river slowed as it grew wider. All Summer long the drop of water watched with awe the sights above and around her: the tugboats and barges and the large merchant ships carrying cargo to the port. One day she tasted a familiar taste which reminded her of her childhood, of the days when she wondered if she would ever see the wide world.

It was evening when she reached the mouth of the river and was swept out to sea. Above her, silver Mother Moon shone softly and the stars twinkled merrily. Below her, Mother Ocean's currents swirled gently, carrying the drop of water on her way. All was peaceful and quiet. She was home again.

These three stories do use not much anthropomorphism because I thought that students would easily relate to the characters and situations without portraying their human characteristics too explicitly. Nature stories that have a moral element or pedagogical stories benefit from anthropomorphism, however, because the characters' human qualities and behaviors help the students relate to the situations and themes more easily.

Second Grade Stories

In second grade, stories of Home Surroundings can portray specific aspects of the natural world and depict natural processes and cycles. Such stories provide an imaginative foundation for geography and natural and physical science classes in the middle and upper elementary grades. Young children today have been stuffed full of indigestible explanations and concepts, but they are still receptive to the depiction of nature with human qualities.

Journey to the Sea

This story takes up the theme of how a river forms from its tributary. It echoes some of the drop of water's journey but from the point of view of the flowing water. To make the story more direct and engaging, Grandfather River speaks in the first person and his listeners interrupt him with questions, as children will. In the first part of this story, I repeated the refrain "No," he answered three times. This kind of repetition enhances the children's participation in the story because they know what is coming (until they don't).

Once upon a time, there was a river that flowed all the way to the sea. The river had many children and grandchildren—creeks, streams, and brooks that came from high up the mountains—as well as other relatives—small rivers that joined him after flowing through fertile valleys.

One day, one of the streams said, "Grandpa—tell us your story."

"What do you want to know?" asked Grandpa River.

"Well, you weren't always so grand as you are now, were you?" asked the stream.

"No," he answered, "not at all."

"And you weren't always so slow as you are now, were you?" asked a brook.

"No," he answered, "not at all. I was once as fast and as spritely as you."

"And you weren't always as deep as you are now, were you?" asked a creek.

"No," he answered, "not at all. I once ran over the rocks and pebbles just like you, and I delighted in making rainbows as my water flew high from the waterfalls."

"So tell us!" they said. "Tell us your story."

"All right," answered Grandpa River. "I will."

"I was born high on the mountainside," said Grandpa River, "born from the melting Winter snow and Spring rains. In those late Winter nights, it was cold, and sometimes I froze a bit, but during the warm days I melted and continued to run downhill, ever down. As I ran quickly downhill, I carved my way through the rocky soil, sweeping around the boulders and stones and tree trunks.

"When I was very young, I often dried up in the Summertime. The mud of my bed grew hard and cracked, and the woodland creatures that used to come to my banks to drink had to find their water elsewhere. But in the Autumn I flowed quickly again, making my merry way, and every Spring I gushed with new water.

"Year by year I made it further down the mountain, and as I grew larger and longer, I carried more water and dashed with more force against the boulders. Sometimes I dislodged them and they would come tumbling down my banks. Now my current carried smaller rocks and pebbles ever downstream, and they scoured my bed, making it smooth in places.

"As the years passed, I grew even larger and swifter and took my mountain carving ever more seriously. Now I was determined to create a large, deep channel for myself, and I pushed my way through the rocky soils, uprooting trees as I passed. I had grown wider now, and the deer that came to drink from my waters could no longer jump across me as they once had. Year by year I worked and worked, until the path I made down the mountainside became broad enough to be called a valley.

"By now I had come a very long way and flowed far down the mountain, through the valleys, to the plains below. Although I still raced down the mountainside, once I got to the lowlands, I slowed down. I had to conserve my strength for my long journey to the sea, and I carried so much water that I couldn't move as quickly. Every day, I thought about Mother Ocean, and I couldn't wait for the day when I would meet her."

"Where did all the water you gathered come from?" asked a stream.

"Ah, that is the joy of it," said Grandpa River. "Even when I was young, just a little older than you, I began to join up with my brothers and sisters who were also flowing down the mountainside. It was always a happy occasion when we met farther downhill and would flow together. Year by year, more and more relatives joined me and my brothers and sisters as we made our way. We call those relatives my tributaries, and it certainly feels to me that all of them pay tribute to their parents with the water they give."

"And have you made it to the sea?" asked a brook.

"Oh, yes," said Grandpa River. "I still remember how happy I was when I saw Mother Ocean spreading her vast reaches before me. As I joined her, I felt like I had come home, home to the source of all being. In that home there was nothing left for me to do, just to rest and feel glad for all that I had seen and done during the long years of my life. And I rejoiced that I would keep flowing back to my Mother forever."

"Thank you, Grandpa, for telling us your story," said the streams.

"Yes, thank you, Grandpa!" said the brooks.

"Thank you!" said the creeks.

"You are welcome," said Grandpa River. "And I hope that one day you will have children and grandchildren of your own, and who knows, perhaps they will ask you to tell them your story."

How the Seeds Were Scattered, Part 1

This story describes the many ways that seeds are scattered to find new homes. It was my first attempt at an origin story. What inspired me was the question, "How come there are some places where different types of wildflowers grow all mixed together and other places where they grow only among their own kind?" Rather than seek a scientific answer to this question, I let my fancy take me where it would, and this story rose up in my soul.

After I had told the firsts part, I was left with a feeling that there was more to say. What about the seeds of the trees and shrubs? And not all seeds are scattered by the wind. How could I account for that? The second part of the story wasn't too hard to compose because I had already set the stage with the first part.

A note about the characters: In this story I decided use the temperaments to help me bring the characters to life. Father Sun was portrayed as a phlegmatic who is not easily stirred to action; Mother Earth was portrayed as a melancholic who cares deeply about everyone and everything; and Brother Wind as a choleric who springs into action and takes charge.

> When we walk by a farm, we see neat gardens of vegetables, fields of grains all planted in straight furrows, and orderly orchards with their rows of trees. Each plant has its place among others of its kind. How is it then, when we walk through the woodlands and meadows, that the plants grow all together, mixed together, topsy-turvy? It wasn't always this way...
>
> Long, long ago, when the earth was young, Mother Nature's garden was neater, more orderly than the best-kept farm. In the forests, trees were planted in neat stands. Oaks grew in great groves; pines soared skyward in straight rows; there were stands of maple, beech, spruce, and hemlock, each growing among its own kind. The forest clearings were lined with borders of ferns and mosses with bunches of sweet violets at each corner.
>
> The meadows too had been carefully planned: In one only goldenrod grew, in another only chicory; there were meadows of daisies stretching in a sea of white and others of asters rippling in purple waves. The byways were lined with blackberry blossoms, and here and there were clumps of wild roses. Every place on the earth displayed Mother Nature's love and care. For years, the plants thrived and grew. All was at peace, until...
>
> One day, a restless daisy turned to her neighbor and said, "I wish I could move away! I'm tired of this meadow with its incessant daisy talk! Don't you wish you could see more of the world?"

"Yes," answered her friend, "Much as I like living among my friends and relations, I'm ready for a change, but what can we do?"

"I don't know," answered the first daisy, but I'm not willing to remain here forever."

Some other daisies overheard what they said, and soon the field was a-whisper with talk of moving away, seeing the world, and making new friends.

Mother Nature overheard their talk and was sad. She could not bear for any of her children to be unhappy. "What can be done?" she wondered. "They cannot move. All the plants have their appointed places in my garden. I will ask Father Sun. Perhaps he will give me some good advice."

The next morning Mother Nature went to Father Sun and told him what she had heard. "What can I do?" she asked. "Don't do a thing," Father Sun replied. "Leave them alone and they will soon forget their troubles."

But the daisies didn't forget. They spoke of nothing else, and soon the goldenrod and the chicory in the neighboring fields were also talking of moving away and seeing the world too. Mother Nature overheard their talk and thought, "I must do something. I will ask Sister Rain. Perhaps she will give me some good advice."

That afternoon Mother Nature went to Sister Rain and told her what she had heard. Sister Rain remained lost in thought for a long time. Finally she said, "I'm so sorry to hear that the flowers are unhappy. But what can be done? You can't transplant them all. I'm afraid they will simply have to stay where they are and accept their lot."

The next day, as Mother Nature roamed over the earth, she heard nothing but eager talk of the future and far-off places. "I must do something," she thought. "I will ask Brother Wind. Perhaps he will give me some good advice."

That evening Mother Nature went to Brother Wind and told him what she had heard. Hardly had she finished before Brother Wind said, "I know exactly what to do! Tell the flowers that they can't be transplanted now, but if they wait patiently until next year, their wish will be granted."

"But what will you do?" asked Mother Nature.

"Just wait and see," said Brother Wind. "Wait and see."

All that week Mother Nature roamed the earth, promising the flowers that if they waited patiently until next year, their wishes would be granted. They thanked her gratefully, and the fields were full of joyful talk of what the new year would bring.

Meanwhile, Brother Wind was busy. He summoned his helpers and told them what to do. All Summer long, they watched the flowers carefully. When they dropped their seeds, Brother Wind's helpers caught the seeds and flew over the fields, scattering the seeds here and there. All through the Summer and early Autumn his helpers worked hard. Then Winter came and covered the earth with a blanket of snow.

Finally Spring returned, and Father Sun warmed the earth once again with his bright, warm rays. The seeds sprouted. The green shoots poked their heads above the earth. The plants grew leaves, then buds. As Spring turned into Summer, the buds opened.

"Hooray!" cried the white daisies, when they found themselves scattered among the goldenrod and the blue chicory.

"Hooray!" cried the red poppies, when they found themselves growing among the blue cornflowers and purple asters.

"Hooray!" cried the black-eyed Susans, when they found themselves flowering among the white Queen Anne's lace and the purple lupines.

"Hooray! Hooray! Hooray!" cried all the flowers, as they looked around at the new sights and made new friends.

Mother Nature heard their cries and looked down at the earth. Gone were the neat patches of white, gold, and blue. Gone were the ordered borders of pink and purple. Gone were the winding rows of white and yellow. Instead the fields and meadows, the highways and byways were awash in splashes of color. Mother Nature's heart swelled with joy. She had never seen anything more beautiful. "Next year I will do more!" she promised. "Each year I will do more, and the whole earth shall reflect God's glory."

But that, dear children, is another story.

How the Seeds Were Scattered, Part 2

As Spring turned to Summer Mother Nature marveled at the splendor of the flowers in the fields and meadows. She had never expected that they could look so beautiful all mixed together. She was not the only one that noticed the changes. The bushes and shrubs along the highways and byways noticed and wondered why they could not move away and see new sights. The trees of the forest in their stately groves noticed and wondered why they had to live among their own kind and never meet new friends. It didn't seem fair: If the flowers could move away from their appointed places, couldn't the trees and shrubs be granted the same privilege?

Throughout the Summer Mother Nature overheard their conversations and was sad. She had promised that she would do more, but what more could she do? Already Brother Wind's helpers were complaining; they were not used to such hard work. Because they had to watch the flowers so carefully, they had no time to play in the warm Summer sky. It didn't seem likely that they would be willing to gather and scatter the flower seeds, much less help with the trees and shrubs. But who else could help?

One morning Mother Nature went to Father Sun and told him her problem. "What more can I do?" she asked. "I want to help the trees and shrubs, but I don't know how."

"You've done enough." said Father Sun. "If you start giving into every request, you'll never hear the end of it. Leave the trees and shrubs alone; they'll get over it."

But the trees and shrubs didn't get over it. When they saw how happy the flowers were to be growing in new places, they yearned to live elsewhere. They spoke bitterly about the injustice that had been done, and their leaves and branches drooped. Mother Nature saw how unhappy they were and grew even sadder. "Who else could help?" she wondered.

One afternoon she went to see Sister Rain and described her problem. "What more can I do?" she asked. "I want to help the trees and shrubs, but I don't know how."

Sister Rain thought a long time before answering. "There's nothing to be done," she said sadly. "If Brother Wind can't help you, the trees and shrubs will just have to suffer and accept their lot."

Mother Nature was not satisfied. At last she went to Brother Wind. "I know that I can't ask you to do more," she said, but perhaps you have an idea of what I can do."

Brother Wind didn't answer immediately, but soon his face grew cheery and he said, "I know just what to do. I can help you, but you'll also need some other helpers."

"Tell me!" Mother Nature begged, and Brother Wind told her his plan.

"Last Summer my helpers worked hard," he said. "They had fun and were glad to help, but I don't think that they are willing to do as much this year. So we must find a way to scatter most of the flower seeds without my workers' help.

"But how can we do that?" asked Mother Nature.

"It's easy!" said Brother Wind. "The flowers' seeds must fly to their new homes by themselves."

"By themselves?" asked Mother Nature.

"Yes," said Brother Wind, "and here's how: My helpers will show the plants how to make sure that I can carry their seeds and scatter them with my breath whenever I blow."

"What a wonderful idea!" said Mother Nature. "But do you think your helpers will agree?"

"They will if I ask them to," said Brother Wind.

"But what about the trees and shrubs?" asked Mother Nature. "You haven't even mentioned them."

"I can't help you there," said Brother Wind. Mother Nature's smile left her lips. "But I know who can," continued Brother Wind.

"Who?" asked Mother Nature. Brother Wind whispered something in her ear. Mother Nature's face brightened, and she said, "Of course! I should have thought of them! Thank you, thank you, Brother Wind!"

All Summer long Brother Wind's helpers taught the plants how to prepare their seeds to fly far away. Some plants, like the grasses, made their seeds so small that they were just carried by the wind. Others, like the dandelion or the milkweed, fixed silken hairs onto their seeds to act like parachutes which were carried by even the slightest breeze. Some, like the jewelweed, stored their seeds in pods that were like catapults. If anything brushed against them, the seeds would shoot out and fall far away from the mother plant.

"How ingenious!" said Mother Nature, and she thanked Brother Wind's helpers for all they had done.

But what about the trees and shrubs? Who was to help them? The next day after talking to Brother Wind, Mother Nature called all the animals and birds to her and described her predicament. She asked them to think of ways they could help her, and before long they began to offer suggestions.

"We could help the oaks and nut trees," said the squirrels. "Every Fall we'll gather their acorns and nuts and bury them. Those we don't eat will sprout and grow where we plant them."

"And we could help the berry bushes," said the birds. "We'll eat the berries and carry their seeds to far-off places. Wherever we drop them, they can grow."

"And I have a suggestion for the maple trees," said a cricket. "They can put little wings on their seeds so that the seeds will fly spinning through the air and land far away from their parents."

"And I know what we should do about the burdock!" said a rabbit. "They can put little spiky hairs on their seed pods, and they'll catch on our fur as we're running past. When we pick out the burrs, the seeds will drop in a new place and grow there."

"Splendid!" cried Mother Nature. "Splendid! I knew you would be able to help me."

All Summer long and into Autumn, the animals helped the trees and shrubs. The trees and shrubs stopped grumbling, and everyone was happy to be working together, and in such ingenious ways.

Winter came and covered the earth with a blanket of snow. Then Father Sun shone bright again, and his warm rays melted the snow. The seeds and nuts swelled and sprouted. They thrust their heads above the ground, grew stems and leaves.

"Hooray!" cried the young oak trees as they found themselves growing among the beeches and maples.

"Hooray!" cried the young blackberry bushes as they found themselves growing among the sumac and hawthorn.

"Hooray!" cried all the flowers as they found themselves scattered even farther afield than the year before.

"Hooray! Hooray! Hooray!" cried all the plants as they saw new sights and made new friends.

Mother Nature heard their joyous cries and looked around the earth. Gone was the neat garden she had planted so many years ago. All of the plants and trees, the flowers and shrubs grew in glorious abandon. Gone was the careful order she had planned. It its place was a wild, fresh beauty that she could never have imagined. Her heart swelled with joy. She had kept her promise. The whole earth did reflect God's glory.

When I saw how the students responded to the previous story, I decided to compose a similar story about trees. I wanted to "explain" how trees ended up in different vegetative zones where only a few species flourish. How could I account for that in an imaginative way but stay true to my intention of providing a true foundation for Nature Study and Geography in later grades?

I decided to take the opposite approach to the seeds: The trees would initially begin in the tropics, with many types mixed together, which made them unhappy. Mother Nature's solution is to separate them and to give each kind of tree a suitable home.

As in some of the other stories, I used repetition and parallel development as literary devices that would create a sense of symmetry and completion.

The Trees, Part 1

Long ago when the earth was young, all the trees lived in the hot tropics where Summer never ends. In mighty forests they grew, warmed by Father Sun who shone brightly every day and bathed by Sister Rain whose soft rains fell every evening. They grew in great numbers, all mixed together in their many varieties.

Many of the trees were happy in the rain forest. They liked the animals that lived among them—the lively monkeys, the stealthy panther, the raucous parrots, and the giant pythons. They liked the bright-colored flowers that bloomed in the clearings and the clouds of delicate butterflies that settled on the moist ground.

Not all the trees were content, however. The oaks and maples didn't like the constant noise and bustle of the animals, birds, and insects. The pines didn't like having so many different kinds of trees growing so mixed together that you could hardly find a friend. And the spruces and firs didn't like being so crowded in the moist, clammy heat.

"Let us move away!" called a mighty oak tree to a slender maple one day.

"Let us move away!" cried a lofty pine to its fellow across the clearing.

"Let us move away!" whispered a scrubby spruce to a fir tree growing nearby.

"Let us move away!" called all the trees.

"Away, away," came the echo.

Among the groves of the rainforest walked Mother Nature, listening to the trees. She had thought they would be happy living all together and was sad to hear their complaints. That night a deep sleep fell over the oaks and maples, over the pines and the spruces and firs. When they awoke, they found themselves in places they had never seen before.

The first thing the oaks and maples noticed when they awoke was the quiet. Gone was the constant chattering of monkeys. Gone was the raucous screeching of parrots. Gone the drone of insects. The forest where they now grew was still. As Father Sun rose, the gentle cooing of the mourning doves could be heard. Leaves rustled softly as a chipmunk scampered off to bury a nut, and a majestic stag walked among the trees without a sound.

The first thing the pines noticed when they awoke was their friends on every side. Gone was the confusing variety of trees. Gone the many vines and creepers that grew up the trunks. Gone the forest floor littered with all matter of plant debris. The forest where they now grew had only fellow pines. The ground was covered with needles, and small clumps of ferns grew here and there in the clearings. The air was cool and fresh with a lovely scent.

The first thing the firs and spruces noticed when they awoke was the bracing air. Gone was the damp and mist of the rainforest. Gone was the awful heat. Gone confusion of many trees growing in tangled webs. The forest where they now grew had plenty of space. Each fir grew in her own spot, each spruce had room to spread his branches. The trees could see far over the tundra, where the lonely hawk circled high.

"Thank you!" called the oaks and maples to Mother Nature. "Thank you!" cried the pines. "Thank you, thank you!" whispered the firs and spruces.

Mother Nature heard their voices and smiled. Her children were happy again.

The Trees, Part 2

The months passed, and the oaks and maples were happy. How they relished the quiet of their woodland home! Slowly the Summer days passed and the days grew shorter, the nights cooler. One morning the forest was covered with delicate crystals of frost that glistened like jewels in the dawn's early light. When the oaks looked at the maples, they were astonished at what they saw: The leaves of the maples had turned color. Gone were the drab greens of late Summer. Everywhere leaves were a brilliant yellow, blazing orange, or flaming red. The forest was filled with the fire of the colored leaves. The oaks called to the maples, who looked at their leaves in astonishment. The maples looked at the oak trees, whose leaves had turned brown in the night. The oaks, who hadn't noticed their own new colors rustled in surprise

"How lovely!" cried the oaks.

"How lovely!" the maples answered.

Mother Nature heard their voices and smiled.

The months passed and the pines were happy. They rustled and whispered, talking to their friends in their forest home. It was cold, and the trees shivered as Brother Wind's Winter blasts blew through their branches. The days were short now, the nights long and bitter cold. One night something awakened the pines. The forest was dark and still, but the trees felt a gentle breath wafting through their branches. What could it be? Birds? Insects? The pines looked around but could not see anything in darkness. They could not sleep that night and waited eagerly for morning. The first glimmer of gray dawn revealed a sight the pines had never seen: The forest was covered with soft, white snow. It was everywhere, resting on the branches, nestling among the needles, and covering the ground in a blanket of white. "How lovely!" cried the pine trees. "How lovely!" Mother Nature heard their voices and smiled.

The months passed, and the firs and spruces were happy in the far northlands. The days grew colder and the grasses withered and turned brown. The lakes and ponds began to freeze, and each day the geese flew southward in great flocks, filling the air with their farewells. The nights were long and cold now, and the trees watched the stars sweeping majestically across the sky. Each day Father Sun rose lower on the horizon and made a low, flat arc across the southern sky. Finally the day came when Father Sun did not rise at all. The firs and spruces looked at each other in surprise, wondering whether he would ever come again. Darkness surrounded them, and the stars twinkled brightly.

Then, in the darkness, there was a shimmer, a shimmer of light: yellow, red, blue, and violet. The colors flashed and flickered in dancing veils. The sky was bright now, hung with curtains of light. The firs and spruces had never seen anything more beautiful. "How lovely!" cried the firs.

"How lovely!" echoed the spruces.

Mother Nature heard their voices and smiled. Her children were happy, and the world was filled with splendor.

V.
Touching the Children's Souls
Stories with a Moral Element, Parables, Pedagogical Stories, and Birthday Verses

Some nature stories do more than tell a story—they tell students the meaning that lies within the story. I call these nature stories with a moral message. These stories achieve their purpose by having characters and situations that can be viewed in human terms, so that the lessons the children learn from them are human lessons.

Children are quite at home with the anthropomorphism of nature stories because they perceive everything in the world in human terms. Until about the age of nine, most children do not differentiate themselves from their surroundings, and they do not see the natural world and its creatures objectively and "realistically." Because they view the world with "eyes of spirit" we can convey deep truths about the world clothed in everyday terms.

Rudolf Steiner pointed out this fact in *The Kingdom of Childhood* (1924), when he said:

> Before the ninth or tenth year, the child does not know how to differentiate itself as an ego from its surroundings. Out of a certain instinct children have long been accustomed to speak of themselves as "I," but in truth they really feel themselves within the whole world.... For children, everything is one, and they are also one with their surroundings. Not until the age of nine or ten do children really learn to distinguish themselves from their environment.

> Therefore it is important to speak of everything that is around the children—plants, animals, and even stones—in a way that all these things talk to each other, that they act among themselves like human beings, that they tell each other things, that they love and hate each other. You must learn to use anthropomorphism in the most inventive ways and speak of plants and animals as though they were human. (30-32)

Anthropomorphism allows the storyteller to use the natural world as the stage on which human challenges and conflicts are expressed and resolved. To an adult, such stories may seem transparently unrealistic, but to a child, they portray the world that the child experiences. Nowadays, anthropomorphism is widely rejected because modern adults think that it will lead children to misunderstand the world. As a Waldorf teacher, I disagree. I think that the best way for children to develop a full understanding of the world is to first to unite themselves with the world through their will. Then they should experience the world through their feelings. Children who have had this opportunity will develop a deeper and richer level of understanding when they are ready to approach the world through their thinking.

We can find many nature stories with a moral element in Victorian literature. While a bit old-fashioned, these stories are a treasure trove of storytelling possibilities. Although some Victorian stories are overly religious or sentimental, many others convey moral truths in a language suitable for children. I recommend this genre for teachers who are looking to use nature stories as part of their students' character education. Numerous books are available online, and some are included in Suggested Sources, Appendix 4.

The teacher's original nature stories are the most potent type of story, however, because students experience more deeply anything that comes from their teacher. While teachers may feel daunted by the challenge of creating their own stories, they may well find that if they embark on the endeavor, they will receive help from unseen sources. Specific guidance for writing nature stories is provided in Chapter 7.

Over the years I composed several stories that contain a moral message. I share these in the hope that they will inspire others to take up the challenge.

The Light of Men

This story was told to first graders at the beginning of Advent, a time when the darkness threatens to extinguish hope. The idea for the story was inspired by Eugene Schwartz's story "Why the Setting Sun Turns Red."

> For many months Father Sun had shone down on the earth with all his might. He had warmed the snow until it melted, and the streams ran once again, singing their merry songs. He had warmed the earth until the seeds sprouted and grown and the trees sent out their tender green leaves and dainty blossoms. He had warmed the air until the fruits swelled and ripened. All through the Summer he had beamed down upon the fields of golden corn and orange pumpkins. In the Autumn, the trees reflected his warm gaze in their gold and scarlet leaves. Now he was tired and needed to rest.
>
> Mother Moon saw how tired Father Sun was becoming and how old he looked. In the Spring he had been young and fresh, looking down on the earth with his strong, bright gaze. He had needed so little sleep, and all through the Summer he

had stayed up late into the evening so that he could look over the richness of the earth just a little longer. Now he was always tired. His glance had become duller and his light dimmer. By late afternoon he was ready to sleep.

Mother Moon grew worried and said to Father Sun, "You need to rest or you will never regain your strength by Springtime. Let me shine upon the earth and warm it with my light." With a weary nod Father Sun agreed and thanked her.

Every night Mother Moon shone down on the earth with all her might. In the dark sky she blazed like silver, lighting the way for tired travelers. But Mother Moon was not as strong as Father Sun. Try as she might, she could not make up for his fading light. In the course of a month, she too needed more and more rest. Finally there was but a sliver of her to be seen, and the nights grew darker.

Mother Moon called her sons and daughters to her and said, "Your father is exhausted from his labors last year. I have tried to light the earth, but my strength is not equal to the task. I need your help so that people will not have to live in cold and darkness."

The stars and planets shone as brightly as they could, sparkling like jewels in the dark sky. Each night they swept across the sky, and even in the dim morning light, the brightest of them could be seen. But try as they might, the stars and planets were not able to light the earth. All they could do was guide weary wanderers on their way.

The nights grew longer and darker. As the nights passed, however, the stars began to notice something. They awakened their mother and said, "Look down on the earth. Each night it has become brighter."

Mother Moon saw that the earth was shining with a dim light. "What could it be?" she wondered. "Is it the lamps in the windows or the fires on the hearths?"

It was the dead of Winter. On the longest and darkest night of the year, Father Sun and Mother Moon slept soundly, and the stars swept in their courses. But the earth was not dark. With each passing hour it grew ever more light. The stars awakened their mother. "Look," they said. "The earth is gleaming!"

Mother Moon looked and saw that the earth shone brightly with radiant light. "What can it be?" they asked. "I do not know," she answered. "Let us ask Father Sun."

It was midnight, and Father Sun was awakened from his long sleep. He looked down upon the earth, and indeed, it was shining as bright as day. "From where does this light come?" asked his children. "It is brighter than lamplight and warmer than firelight!"

Father Sun cast his glance far down to earth. "That light is the light of man," he said. In the darkest night, a new sun is born in the human heart."

The Two Kingdoms

This story was written for a class that had many hardworking students as well as some who were reluctant to exert themselves. In it I tried to characterize the difference between a materialistic existence and one that is dignified by work.

In *The Kingdom of Childhood* (1924) Rudolf Steiner said that one of the reasons that it is good for children to have the same teacher for many years is that they can return to stories and themes that they brought at different stages of their education. This story laid a foundation for middle-grades Geography lessons about natural resources and for discussions in the upper grades about the unequal distribution of wealth in the world.

> Once upon a time there were two kingdoms, as different from each other as any two kingdoms can be. One was mountainous and barren, with rocky hills and scorching deserts. The other had grassy plains and great tracts of woodlands, and it was bounded by the sea. Lest you think that the first kingdom was poor, let it be known that untold treasures of gold and silver and precious stones lay hidden deep within the mountains. Miners were hired to work day and night to bring forth the great wealth from the earth, and the people of the kingdom lived in ease and luxury.
>
> In the second kingdom, people had to work hard for a living: Woodcutters cleared the great forests, sending the mighty tree trunks down the rivers to be sawn into wood. Farmers broke the sod and tilled the soil, bringing forth crops by the sweat of their brow. And fishermen worked from dawn to dusk to try to eke out a living from the sea.
>
> One day a rich man from the first kingdom took a pleasure trip and went to visit his poorer cousin, who lived in the second kingdom. When he arrived, he found his cousin in the field, trudging behind a team of oxen, plowing a furrow. The poor man's hands were calloused and hard; his face was wet with sweat; and his boots were caked with clay. Nevertheless he was whistling a merry tune. The rich man hailed his cousin, who stopped his plowing only long enough to tell him to make himself at home and that he would join him in the evening.
>
> That evening the two men sat and spoke together. "How can you live this way?" asked the rich man. "Look at you—you are little better than a peasant! Leave your

farm and come live with me. I shall see to it that you shall not lack for riches."

"What sort of riches are these?" asked the poor man.

"Why, riches of every sort!" answered his cousin.

"Tell me about them," said the poor man.

"Indeed I will," the rich man replied. "In our kingdom, none of the king's subjects need work, for our work is done by hired laborers from neighboring lands. Such wealth lies beneath the ground that our king distributes it among his subjects and still has enough to make him the wealthiest and most powerful ruler on earth."

"And what do you do with your wealth?" asked the poor man.

"Anything I want," answered his cousin. "I have a grand home that lacks for nothing. I eat the finest foods and drink the rarest wines. My servants fulfill my every need, filling my life with pleasure and amusement. I need not sow to reap a rich harvest. What more could a man want?"

"I am not sure such a life appeals to me," said the poor man.

"How could it not appeal to you?" asked his cousin in disbelief.

"I love to work," said the poor man, "and in your land you need not work. Work yields its own rewards. Through work a man comes to know himself. Here I reap what I sow: If I work hard, my family has plenty; if not, we starve. If I look after my animals they thrive; if not, they suffer. If I care for the land, it remains fertile; if not, it grows barren. Through work I create, and thus I feel myself closer to Him who created us all."

That night the rich man lay in bed thinking about what his cousin had said. He could not understand how someone could be content with so little. Yet he sensed that his poor cousin had something that he lacked in himself. "What is it?" he asked himself. "What is it that makes me think that he is far richer than I shall ever be?"

> Work is love made visible.
> And if you cannot work with love but only with distaste,
> it is better that you should leave your work
> and sit at the gate of the temple and
> take alms from those who work with joy.*

---

*From "On Work" from *The Prophet* by Kahlil Gibran

Parables

> To see a World in a Grain of Sand
> And a Heaven in a Wild Flower
> Hold Infinity in the palm of your hand
> And Eternity in an hour
> 	– from "Auguries of Innocence" by William Blake

In his first published work on education, Rudolf Steiner introduced the idea of working with parables of nature. He returned to this idea repeatedly in many of his later educational lectures.

In the following passage from *The Education of the Child in the Light of Anthroposophy* (1907), Rudolf Steiner speaks about the importance of presenting natural laws to students through living pictures rather than intellectual concepts. As an example, he uses the image of the butterfly emerging from the chrysalis as a picture of the soul emerging from the body at death. Rudolf Steiner states that such an image engages the children's feelings, and that later in life, children who have received such parables will approach the problems of existence differently than those who have been fed only dry concepts.

According to Rudolf Steiner, parables for all the laws of nature and the secrets of the world have been embedded in the natural world by the spiritual world. Teachers can discover these parables through their study of spiritual science, and if the teacher believes in the parables, students will be spiritually nourished by hearing them.

> In another connection, for the period between the change of teeth and puberty, it is important to present living pictures—or symbols, as it were—to the mind. It is essential that the secrets of nature, the laws of life, be taught to children, not in dry intellectual concepts, but as far as possible in symbols. Parables of the spiritual connections of things should be brought before the souls of children in such a way that behind the parables they divine and feel, rather than understand intellectually, the underlying law in all existence. "Everything passing is but a parable" must be the maxim guiding all of our education during this time. It is of vast importance for children that they receive the secrets of nature in parables before they are brought before their souls as "natural laws" and so on.
>
> An example may serve to make this clear. Let us imagine that we want to tell a child about the immortality of the soul, of the coming forth of the soul from the body. The way to do this is to use a comparison—for example, the butterfly coming out of the chrysalis. As the butterfly soars

up from the chrysalis, so after death the human soul comes forth from the house of the body. No one can properly understand this fact in intellectual concepts who has not first received it through such a picture. By a parable such as this we speak not just to the intellect but to the feelings of children, to their whole soul. Children who have experienced this will approach the subject with a completely different mood of soul when later it is taught to them in the form of intellectual concepts. It is a very serious matter indeed for anyone who is not first given the ability to approach the problems of existence through feeling. It is therefore essential that educators have at their disposal parables for all the laws of nature and secrets of the world…

When one speaks in parables and pictures, it is not just what is spoken and shown that works on the listener, but a fine spiritual stream that passes from the one to the other, from the one who gives to the one who receives. If the one who tells does not have the warm feeling of belief in the parable, no impression will be made on the one who hears. For true effectiveness, it is essential to believe in one's parables as one does in absolute realities. And this can only be so when one's thought is alive with spiritual knowledge.

Take, for example, the parable we have been speaking of. True students of spiritual science need not torment themselves to get it out. For them it is reality. In the coming forth of the butterfly from the chrysalis, they see at work, on a lower level of being, the very same process that is repeated on a higher level, at a higher stage of development, when the soul comes forth from the body. They believe in it with all their might; and this belief streams, as it were, unseen from speaker to hearer, carrying conviction. Life flows freely, unhindered, back and forth from teacher to pupil. But for this it is necessary that teachers draw from the full fountain of spiritual knowledge. Their words, everything that comes from them, must have feeling, warmth, and color that comes from a truly spiritual-scientific way of thought.

A wonderful prospect is thus opened throughout the field of education. If it will only let itself be enriched from the well of life that spiritual science contains, education will also be filled with life and understanding. There will no longer be the groping so prevalent now. All art and practice of education that does not continually receive fresh nourishment from roots such as these is dry and dead. The spiritual-scientific knowledge has appropriate parables for all the secrets of the world—pictures taken from the very being of the things, pictures not made first by human beings, but put in place by the forces of the world within things themselves, through the very act of their creation. Therefore this spiritual knowledge must form the living basis for the whole art of education. (26-28)

In later educational lectures, Rudolf Steiner re-emphasized the importance of working with parables and the importance of the teacher's belief in them.

In *Practical Advice to Teachers* (1919) he said:

> If you yourself do not believe that the butterfly is an image of the human soul, you cannot accomplish much with children by using this analogy. You should not allow yourself the false notion that this whole idea is merely a contrived comparison, which it is not; it is a fact presented to us by the divine cosmic order. These things are not invented by the intellect, and if our attitude toward such matters is correct, we come to trust the fact that all nature offers us analogies for the realities of soul and spirit.
>
> As we unite with what we teach children, the way we work affects their whole being. When we can no longer feel with children and instead offer only rational translations of everything that we ourselves do not believe in, we cease to teach children very much. Our relationship to reality must be such that, out of our own comprehension, we bring to children's souls more than an arbitrary picture of the butterfly emerging from the chrysalis, for example, and instead present something we ourselves understand and believe in as given by divine cosmic powers. We must not offer children understanding merely for their ears, but we must communicate from soul to soul. If you remember this, you will make progress. (15-16)

In one of his last educational lectures, Rudolf Steiner returned to this theme, adding:

> As students of spiritual science, we know that an emerging butterfly is a true image of the immortal human soul, which is placed into the world by the gods. We have to imagine that the gods inscribed this picture into the world—that is, the emerging butterfly being an image of the human soul's immortality. We see the higher processes abstracted in all the lower stages of the process. An imponderable relationship arises between the teacher and the students; and the children make real progress in their education as long as you do not get the idea that they are ignorant and you are clever.

> You must stand before the children, aware that this is a fact in the world, and that you are leading them to believe in something that you yourself believe with all your heart. This is how moral imponderables continually enter the educational relationship. This is crucial. (*Human Values in Education*, 55)

Rudolf Steiner did not speak about parables specifically in the context of Home Surroundings or nature study. Instead, we can infer that he intended them to be used in conversations with the students about important topics. However, parables do have a special place in the Home Surroundings curriculum because they create a living foundation for the study of natural science in later years.

Parables are especially valuable in helping students connect more fully with the natural world. They deepen the teacher's view of the world as a revelation of the spirit, so that what is shared with the students is infused with meaning. Most importantly, working with parables strengthens the relationship between teachers and students on the soul and spiritual levels.

I did not compose specific parables for my students; rather, I let some of the parables I perceived shine through the stories they heard and the verses they recited. At times, I brought parables as metaphors in conversations with my students, and in the upper grades, I sometimes hinted at the parables that underlie physical and chemical phenomena.

Mostly, parables helped me as I sought to find meaning in subjects such as physics and chemistry which were so foreign to me initially. In my work with upper-grades teachers over the years, I have encouraged them to look for parables in every subject they teach. Home Surroundings in the first two grades is a wonderful training ground to heighten the capacity to perceive parables in the world around us.

To give you a sense of what kinds of parables you might find, here are two examples from the study of chemistry in seventh grade:

> We can view combustion as a picture of what happens to the human being at the portal of death. When something combusts, it yields ash, gases, heat, and light. Viewed as a parable, we can view the ash as what remains of the physical body, the gases as an expression of the soul, and the heat and light as embodiments of the spirit. We are beings of body, soul, and spirit; at death, these three aspects of our being separate and return to their proper realms.

> We can view a chemical solution as a parable of human incarnation. We can think of the human being as a solute that is dissolved in the solution

of spirit. When we are in the spiritual world between death and rebirth, the parts of our being are disparate, like ions with their positive and negative charges. To re-enter the physical world, we need to precipitate out of the spirit, and in taking on a physical form, we unite spirit, soul, and body.

Although thoughts should not be shared directly with our students, pondering these kinds of parables can help us recognize and appreciate the cosmic mysteries that lie embedded in the mundane details of everyday life. I offer the following verse by Rudolf Steiner to support you in your contemplative work with parables:

> Seek in your own being
> And you will find the world.
> Seek in the worldwide being
> And you will find yourself.
>
> Note the constant swing
> Between self and world
> And you will find revealed
> The human cosmic Being,
> The cosmic Human Being.

Pedagogical Stories

Pedagogical stories are one of the most subtle tools that teachers can use to help students learn a lesson or develop a character trait. They can be composed for a particular student, a group of students, or the whole class. Characters and situations from nature lend themselves especially well to pedagogical stories because they mirror the child(ren) for whom the story is composed.

Teachers who use pedagogical stories recognize their power not only for the child or children for whom they are composed, but for the whole class. The class community is incredibly perceptive about each member's challenges and striving, and they are reassured to learn that the teacher recognizes and understands the situations and is working subtly and artistically to provide support to those in need.

Teachers are often reluctant initially to use pedagogical stories for fear that the story might draw public attention to a child's challenges, thereby shaming the child. I have never seen this happen. First and second graders are remarkably accepting of any efforts to address difficulties, and they are finely attuned to the teacher's positive intentions. The child for whom the story is composed usually unconsciously recognizes that the teacher is working on his or her behalf. Sometimes the child lets us know this in surprising ways.

I once had a student who found it difficult to accept direction and correction, so I composed a pedagogical story about a young colt who rebelled against all attempts to train him. The colt's master eventually won his confidence by being kind and patient and by reassuring the colt that he knew from his own experiences how hard it is to obey. When I finished telling the story, the student for whom I had composed the story blurted out: "I know that story!"

I encourage you to use pedagogical stories as one of the ways to serve your students in their journey of becoming. No matter how inept you may feel, your efforts on the child's behalf will help the child, and many of the child's classmates will offer more inner support to their classmate.

Suggestions about how to write pedagogical stories are provided in Chapter 7.

The Nettle

I had a student in my first class who was difficult to get along with. She took everything personally. She was defensive and defiant. She struck back at imagined insults. Her classmates and I often tip-toed around her to try to avoid any unpleasantness.

This child had many wonderful qualities—she was intelligent, artistic, and athletic; she was truthful, brave, and determined. But she didn't feel that anyone noticed these qualities. She thought that all anyone noticed were her faults.

I wrote this story to reassure this child that even a prickly plant can find her place and be valued for her special qualities.

> On the banks of a slow-moving river, grew a young plant. She had a sturdy stalk, graceful branches, and dark green leaves, but her leaves looked almost gray, for they were covered with tiny hairs. Although she did not look particularly remarkable, there was something different about her. No grasses or flowers grew near her. No birds alighted on her branches; no animals burrowed by her stalk. For any creature that touched her leaves was stung by the poison contained in her leaves' tiny hairs. Thus, although she lived surrounded by other plants and creatures, she felt utterly alone. "It's her own fault that she's lonely," said the other plants. "She's so prickly and stand-offish."
>
> One day, a young man and his lady-friend came up the river with a picnic basket and a blanket in the bottom of their punt. The man steered toward the shore, leapt out of the boat, and clambered up the bank. He had seen a bunch of daisies growing in a clearing and wanted to pick them for his lady-friend. As he came near the stinging nettle, he slipped in the sandy earth, and his arm brushed against the

nettle's leaves. "Ow!" he exclaimed, as he rubbed his stinging arm, which was quickly growing red. "What an ugly and useless plant!"

"It serves him right!" thought the nettle in a huff. "He should have been more careful. That will teach him not to come near me!" But as the day wore on, the man's words echoed in her soul, "What an ugly and useless plant!"

Summer turned to Fall, and the nettle's leaves turned a bright, fiery red, but her soul remained dark and mournful. One day she heard voices of someone coming up the river. "I think it's by this bank over there," said the lay-brother to the monk in the front of the punt. The monk got out of the boat, shovel in hand. "Ah, isn't she a beauty!" he exclaimed. "I'm so glad we found her!"

Carefully, the monk dug around the nettle's roots, making sure not to injure them, wrapped them in burlap, and carried the nettle gently in his gloved hands to the punt. As she floated down the stream, the nettle felt herself becoming drowsy, and listening to the sound of the water lapping against the side of the boat, she fell into a deep sleep.

The following Spring she awoke and looked around. Where was she? She saw the high walls of the monastery herb garden and heard the sound of the bells calling the monks to prayer. Basking in the warmth of the morning sun, she heard footsteps approach and then the gentle voice of the monk. "Ah, there you are, my beauty!" he said, "You have awakened from your long Winter's sleep. Soon your buds will open once more, and we shall brew your leaves into a bracing Spring tonic that will cleanse and fire our blood."

In the monastery garden, the nettle grew and thrived, and she felt at home among the other herbs and flowers. No longer was she lonely, for she had found a place where she was cherished and known. The monks affectionately called her "fireweed," and every Spring, they made a tonic from her leaves that gave them renewed strength for learning and for working.

Birthday Verses

Pedagogical stories or parables can be condensed into birthday verses that use images from the natural world to evoke human truths. Such verses are very nourishing for children because they give them a sense that they are truly known. They also deepen children's relationship to the natural world by making personally significant what might be considered mundane. Birthday verses derive their power from repetition, so a pedagogical message can resonate in the child's soul throughout the year.

Birthday or report verses are most effective if they are composed by the teacher for a specific student, but they can still be very potent if children recite verses that you find. What is most important for birthday verses is the activity of reflecting on the child and finding the right image and message. Chapter 7 provides guidance for composing and using birthday/report verses, and examples for emulation or adaptation—or simply to use as written—are provided in Appendix 3.

VI.
Composing Descriptive Nature Stories

Theory

Rudolf Steiner challenges teachers to transform education into an art. That means to become an artist in the very way we educate children and also to cultivate our own artistic capacities. In Home Surroundings, one opportunity to develop our artistry is to compose our own nature stories. Such stories will have a great impact on our students because they come from us. And the act of creating them will ignite our enthusiasm and deepen the satisfaction that we take in our work.

Rudolf Steiner was adamant about the need for teachers to compose their own stories. In *Soul Economy and Waldorf Education* (1921) he said:

> In this context, something else of great significance must also be considered. What would lazy teachers do when faced with such an educational challenge? They would most likely go to a library and look for books of legends, animal stories, and other similar subjects, then they would read through them for use in the classroom. Of course, sometimes you have to make do with inferior arrangements, but this method is far from ideal. Ideally, teachers would prepare themselves so well for this task—which does require thorough preparation—that a conversation between plants, or a fairy tale about a lily and a rose, comes to children as the teacher's own creation. And ideally, a conversation between the sun and moon should be a product of the teacher's unique imagination. Why should it work this way?
>
> Let me answer with an image. If you tell students what you found in books—no matter how lively you may be—if you tell them what you have read and perhaps even memorized, you will talk to them like a dry and desiccated person, as though you did not have a living skin but were covered with parchment, for there are always death-like traces in one's own being of what was thus learned from the past. If, on the other hand, you are creative in your work as a teacher, your material will radiate with

growing forces, it will be fresh and alive, and this is what feeds the souls of children.

If as a teacher you want to reach children at this age, there has to be a creative urge to clothe the world of plants and animals and the sun and moon with living stories. Once you have engrossed yourself in such imaginative work (which demands a great deal of inner effort), you will hurry to school with steps betraying your eagerness to share these offerings with your class, and the effects of such an endeavor will be wholesome for all the children. Such teachers know very well that their story will remain incomplete until they have seen the radiant faces of those young listeners. (161-163)

Composing nature stories seems a tall order, yet it can be done, and it is well worth the effort. When you attempt to write your own stories, you will find that the creative process makes us more receptive and alive to the world, and, as Rudolf Steiner said, "It is aliveness, that must be the guiding principle. 'Aliveness' in the teacher must pass over to 'aliveness' in the children." (*The Kingdom of Childhood*, 115)

No matter how limited we believe ourselves to be, our creative striving brings out the best in us. It also brings us closer to our students, for they perceive the person who we are trying to become, and they recognize that we are working out of our higher selves to serve them.

Finally, as Rudolf Steiner said in the passage above, the works that we create are much more nourishing to our students' souls that anything from another source that we could "warm up."

I recognize that such creative endeavors take time and energy, and that it is important to remain realistic about how much we can take on. But I strongly encourage all teachers to compose at least some of the nature stories that have moral meaning, and when they cannot, to seek carefully for the stories that best meet their students' needs.

Practice

Nature stories that describe the characteristics of the world and its creatures are the easiest type of nature story to compose. Such stories are similar to the presentations we make in grades 3–8 when we share vivid pictorial descriptions or narrations.

The first step to writing a descriptive nature story is to inform yourself about the subject. This kind of preparation can be done from books and articles, but looking at pictures can also inform us and stimulate our imagination. I encourage you to look primarily at illustrations and paintings rather than photographs because an artist is often able to express some of the subject's essential nature.

In addition to your research about the subject, if it's possible, try to spend some time in direct experience with the subject of your story. This will allow you to make your own observations and form your own impressions, and it will allow the subject to speak to you directly. When observing and reflecting, try to push all of your prior knowledge and preconceptions to the background so that you can be truly present with the subject you are striving to get to know.

Once you have completed your research and observations, you will need to reflect on what you have learned and discovered and begin to separate the essential from the non-essential to give your story focus and coherence. I usually create an outline of the major events in the story so that I can check to see if all the important points are included.

I like to compose stories by writing them out first. Telling a story on paper (or on the screen) allows me to see the story unfolding. I like the security of being able to practice the story by reading it out loud and then rehearsing it silently so that when I tell it, I feel secure in the plot and descriptions.

Some people do not write out their stories but rehearse them in their mind or out loud. And some people wait to tell the story directly to their students without any rehearsal. Although I lack the courage to use this approach, I know teachers who prefer to let the story unfold organically in the moment of telling. While you are new to the process of composing your own stories, try the different approaches and you will quickly discover which works best for you. Below is an example of how I worked with on descriptive story for second graders. Having taught in the upper grades for many years, I had the idea that it would be good to set the stage for our future sixth-grade study of the mineral world by telling stories about rocks. I composed a story about several different stones meeting each other on the bed of a stream and telling each other their life stories. Then I wanted to present the rock cycle in story form.

I decided to use a grain of sand that was metamorphosed into a quartz crystal and then eroded until it returned to its original form. The plot for this story was established by the rock cycle sequence. And I thought that the grain of sand did not need to be characterized fully; it was sufficient that it had gone through an amazing journey.

The impact of a story for young children comes from the vividness of the pictures that the story presents. I am, by nature, not a pictorial person, but teaching demanded that I cultivate my ability to speak in pictures. Composing simple stories such as this one gave me opportunities to develop my imagination and my ability to express it.

A Grain of Sand

On the bottom of the ocean lay a grain of sand, sparkling like a jewel in the faint sunlight. Once before he had lain on the ocean bottom, but that had been years before, many thousands of years before. How he returned to the same place had been quite an adventure.

Long ago, a grain of sand lay on the ocean bottom, amid countless other grains. Every day the river washed new grains of sand into the sea, and they floated out over the bottom, gently settling into their sandy beds. Over the years and centuries, these beds became deep and thick, and the grain of sand was covered by many hundreds of feet of sandy sediment. The weight of these sediments pushed down upon him with great pressure. Slowly he felt himself bonding together with other grains of sand until they formed sheets of rock called sandstone.

One day the grain of sand realized that he was slowly moving downward. The movement was so slow and gentle that he hardly felt it, but when he paid careful attention, he could tell that he was moving. One day the movement grew greater. The ocean floor was shaking, and the beds of sandstone cracked and splintered. When the shaking stopped that day, there were smaller tremors during the next few weeks. The movement continued, however, until the grain of sand was buried under ever deeper layers of the ocean floor.

It was growing hotter where the grain of sand lay, and in the heat, he felt himself growing brighter and clearer. As he looked about, he saw that the entire bed of sandstone was changing from dullish red and brown into shining white quartzite.

As the temperature grew hotter and the movement downward continued, the grain of sand was sinking ever farther under the crust of the earth. He passed other layers of rock—slate, marble, and then granite—as he moved downward. Now the heat was fierce, and the grain of sand felt himself melting. He was no longer solid rock, but a clear sticky fluid that seeped through the deeper layers of the earth's mantle.

One day the grain of sand realized that he had stopped moving downward, and now he lay in a great lake of liquid rock. The lake was hot, and every so often bubbles of gas swelled and burst from its depths. Now the liquid rock began to move upward slowly, back through the cracks and crevices of the layers of bedrock. Sometimes the movement was slow and gradual; sometimes the rock suddenly gave way and a torrent of molten rock rushed into the empty space. Ever upward traveled the grain of sand until finally he came to rest in a cave.

The temperature was cooler here, and the liquid rock next to the wall of the cave began to harden. One day the grain of sand felt himself begin to harden back into a solid form again. Together with countless other grains of sand, he was now becoming part of a quartz crystal. As the liquid rock hardened over the years, the inside of the cave became completely encrusted with quartz crystals. What a pity that the sun could not shine into its depths and be reflected off their many smooth facets!

Although the grain of sand had come to rest, his bed was not completely quiet. Every so often the earth shook a little, and pieces of the crystals would break off. Every Winter and Spring, as the earth froze and then thawed, the cave shook ever so slightly. Soon the roots of trees began to creep down into the roof of the cave, breaking up the rocky walls.

One day the grain of sand saw the sunlight again. A mighty earthquake had shaken the ground, and the mudslides had ripped the trees off the mountainside. The cave had split open and the bright sun shone in. Now wind and water swept into the cave, slowly weathering the rocks. Animals used the cave for shelter, and their rough scratching dislodged some stones, including the one containing the grain of sand.

Out onto the open mountainside his stone was pushed, exposed to sun and wind. The following Spring, with the grain of sand tucked securely inside, it was carried quickly down the slopes in a rushing stream of melted snow. Down the streambed it tumbled, bumping into other rocks all the way.

Years passed, and the rock was carried ever farther downstream. It had grown smaller and smoother now. Its bright white crystals no longer shone, for its surface was all scratched by the many bumps it had received. Ever farther down the stream it was carried, sometimes rushing in a flood, sometimes settling for a whole Summer during a drought. Ever smaller it became: from a rock to a pebble, and finally to the little grain of sand again.

The stream was no longer so rough, for it had become a mighty river carrying its load of silt and sand to the sea. Mixed in was the tiny grain of sand. One day he sand felt the current slacken. Slowly he drifted to the bottom. The water was no longer fresh, but salty. He had reached the ocean!

In the course of the next months, the gentle ocean currents carried the grain of sand farther out to sea. At last he came to rest in a great sandy bed, the place where his great journey had begun. That journey had carried him to the depths of the earth and to the heights of mountains. He had been pushed and pressed, hurried and heated, but always he had remained his own shining self.

Here is how I worked on another story for second grade. I wanted to share a story about contrasting characters, so I chose an oak leaf and a snowflake. I'm not sure exactly how I made this choice, whether it was deliberate or serendipitous. I know that I like strong contrasts (and think they stimulate children's imaginations), and these characters were almost opposites: one being related to the earth, the other to the heavens; one being tenacious, the other more frivolous; one who had always been rooted in one place, the other who had traveled.

I decided that I would let each character tell their life story up to the point where they met. Then they would speak about what lay ahead on their journey together. To make the narrative a little livelier, I had the character who was listening ask some questions and make some comments. I didn't feel the need to develop the characters' personalities in the story. It seemed sufficient to allow their different journeys to highlight their differences. It wasn't clear to me how the story should end, but I finally I decided that once the two characters had formed a relationship through their conversation, they would take the next step of their journey together.

Down to Earth

At the edge of a field grew a mighty oak. It had a wide trunk, sturdy limbs, and broad branches. From the branches hung just a few brown leaves. Oak leaves are stubborn—they don't let go easily—and you can often see a stately oak trees still holding on to some of last year's leaves even while it starts showing its new leaves in the Spring.

This night the sky was dark with clouds. The stars and moon were hidden, and only a bit of their soft light shone upon the earth. Everything was still and very cold. It was December, and the first snow had not yet fallen.

In the middle of the night, one of the oak leaves felt something fall on her. What was it? Was it a bug? Unlikely—because bugs don't come out on cold Winter nights. Was it a bird? Impossible—it was much too lightweight.

"Who are you?" asked the leaf.

"A snowflake," answered her visitor.

"Really?" asked the oak leaf. "I've never met a snowflake."

"Well, I'm pleased to make your acquaintance," said the snowflake. "I've never met an oak leaf."

"Where did you come from?" asked the leaf.

"High, high up in the sky," said the snowflake.

"Tell me about yourself," said the oak leaf. "Tell me the story of your journey down here to meet me."

"Alright," said the snowflake. "I will."

"Believe it or not, I started out as a tiny speck of sand from a beach far away from here. One day, the wind blew me off the beach and lifted me higher and higher, until I was so high above the earth that I could hardly see it. As I was floating, I felt something form on my surface. To my surprise it was a film of water. Day by day more water formed on my until I was covered with it like a wet coat.

"One night, when it got very cold, I felt myself changing. I looked down at myself and found that I had sprouted arms."

"Six of them, right?" said the oak leaf. "I can see all six."

"Yes," answered the snowflake. "That is the sign of my heritage. Every snowflake in my family has six arms. I know there are other families that have different shapes—plates or columns, or furry spikes—but everyone in my family has six arms like me."

"What happened next?" asked the oak leaf.

"I lived up in the wispy clouds for a couple of days, and then, as I got bigger and heavier, I felt myself falling. Down and down I fell. But then a gust of air would carry me back up to the top of the cloud and I would start to fall again. It was fun being carried up and falling down, and my brothers and sisters and I were delighted by the wild ride."

"And then?" asked the oak leaf.

"Finally, I drifted down to earth," said the snowflake, "and came to rest here on you."

"That's quite a journey," said the oak leaf.

"Yes it was," said the snowflake. "And how about you? How did you come to be here so high in this tree?"

"Oh," said the oak leaf, "my journey was very different from yours."

"Well, tell me," said the snowflake. "I'd like to hear."

"I was born in the early Spring," said the oak leaf, "though my baby bud was already on the branch the Autumn before. All Winter I slept, waiting for Father

Sun to warm the earth and to wake up us leaves. Finally, I burst out of my bud and was delighted with my pretty light green color. I grew larger in the warm Spring sunlight, and slowly I turned darker, until I was a beautiful emerald green. All Summer long I looked down on the earth and marveled at everything that was happening. As Summer gave way to Fall, I felt myself changing again: My green faded and I turned a rich golden brown. Day by day, as Brother Wind blew, my brother and sister leaves let go of their branches, until finally, there were only a few of us left. I've hung on to my twig because I wanted to see what Winter is like and to see snow. And now that I've met you, I feel glad that I waited to go back to the earth."

"How will that happen?" asked the snowflake.

"Once I drop down to the ground, I will hug Mother Earth tight. She will fold me further in her arms until I become one with her. And how about you?" asked the leaf.

"I will wait until I fall down to earth, and then wait until a warm day melts me back into water. Then I will seep into the ground and who knows—perhaps I'll find my way to an underground stream and make it all the way to the ocean. Or perhaps I will be soaked up by a plant or tree. Or maybe an animal will drink me. Whatever my path, I will return to Father Sky once more."

"Well," said the oak leaf, "shall we go down together?"

"Yes," answered the snowflake. "Let's do."

When she felt the next gust of wind, the oak leaf let go of her twig and drifted down to earth, the snowflake balanced carefully on her back. There they lay happily together, waiting for the next step on their journey.

VII.
Composing Stories with a Moral Element

Stories that have a moral element can be harder to compose than descriptive stories. Whereas descriptive stories are based on facts, stories with a moral element require us to find a message in some aspect of the natural world. It takes effort to find a message in nature, but our efforts will deepen our students'—and our own—relationship to the world around us.

It is difficult to decide to write such a story in the abstract. It is much more likely that we will be inspired to compose one when we are confronted with a moral question about our students or the class. If we pose that question in the evening as part of our meditation, the images of a story may come to us in the morning or at a time when we do not expect them.

A story with a moral theme need not be profound. It can simply hint at a truth and still perform its task. The main challenge of writing such stories is to find a picture from nature that is analogous to a situation or relationship in the children's lives. Once you have found a picture, you may find that the storyline will suggest itself.

Here is how I worked with a story with a moral element:

My second grade class very large, and I was worried that some of my students were feeling unrecognized. I had the idea that if everyone was assigned an essential task, then they would feel more appreciated. I searched around for a situation in nature which might have a family feeling, with a lot of variety in the context of close relationships. I came up with the idea of the many forms of precipitation. I pondered how to characterize them and decided to try matching the forms of precipitation with the temperaments and with the seasons.

> Hail – choleric – Summer
> Mist – melancholic – Autumn
> Frost and Flake – sanguine – Winter
> Dew – phlegmatic – Spring

But what would be the conflict? In trying to work all together, each character's temperament was not an asset; rather, it created difficulties. The solution? Have the characters perform tasks suitable to their temperaments. That would give them each a chance to put their temperament to use in the performance of their task and would help them feel successful and appreciated.

My second graders were used to hearing about how Mother Nature was so wise that she always knew what her children and grandchildren needed. I had her appear repeatedly in my stories because I wanted my students to look upon nature as the source of answers to human problems.

The Children of the Rain

It happened once long ago that Sister Rain went to Mother Nature with a problem. "It's my children," said Sister Rain. I just don't know what to do with them anymore."

"What do you mean?" asked Mother Nature.

"It's hard to describe," said Sister Rain.

"Try anyway," said Mother Nature. "I'm sure it will become clear."

"Take my son Hail, for instance" said Sister Rain. "He's so wild, always racing around and throwing himself into things. He doesn't even know how destructive he is. Last week he flattened a whole garden of beautiful flowers and he didn't even notice. He was so anxious to show that he was the fastest of us all.

"Then there's my daughter, Mist," continued Sister Rain. She's so dreamy, she can't remember a thing. I tell her to follow me, and first thing I know she's gone, drifted into a hollow to look at some ferns or dawdling under a bridge, admiring the little pools.

"My twins are just as exasperating," Sister Rain continued. "Flake loves to dance, and, I must admit, she's very good. But ask her to water the fields and she flits around here and there. Her brother Frost is a little more reliable, but he can't resist a chance to paint everything with his brush. By the time he's done, I'm way behind with my chores.

"And then there's Dew, bless her heart. She's so thorough and picky that she never finishes any of her chores. Every blade of grass must glisten just so, every strand of a spider web has to have the exact same number of drops. It's a wonder I get anything done when she's around.

"Do you have any ideas what I can do?" asked Sister Rain.

Mother Nature smiled her mysterious smile. "Yes," she answered. "But I want to talk to your children first."

That afternoon Mother Nature spoke with Hail, Mist, Flake and Frost, and Dew. They were eager to hear what their grandmother had to say, for they knew that their mother had been irritable and unhappy lately, and they didn't know what to do about it. Mother Nature told them what she had in mind and asked them if they would do what she asked. Gladly they agreed to her plan, for it made perfect sense to them.

While Mother Nature was talking to her grandchildren, Sister Rain waited patiently. She was curious about the pleased and mysterious looks her children gave her as they left. "Well, what was that all about?" asked Sister Rain.

"Here's what I suggest," said Mother Nature. "I've discussed it with your children, and they are amenable.

"You have a great deal of work to do, my daughter, and it's clear you need help, but the right kind of help. When you described your children to me yesterday, you gave me an idea. Each of your children has special talents, but you haven't found a way to use them. They are eager to help you, but they don't seem to know what's most important or how they can be most useful. What's needed is that each one does what he or she does best, so here's what I suggest:

"In the Summertime, let Hail help you. You know he loves to race after his uncle Thunder and dash about with his cousin Lightning. Put his energy to use when you need to water the dry and thirsty fields.

"In the Fall, let Mist help you. She will wrap herself around the golden leaves without disturbing them, gently watering the Autumn plants. If she forgets or loses herself in her thoughts, there's nothing to worry about, for the plants do not need so much tending at that time.

"As for Flake and Frost, let them be your Winter helpers. On the coldest days Flake won't mind dancing through the hills and fields, dusting them with snow. And on those clear, cold Winter nights, Frost can paint every window pane to his heart's delight.

"In Spring, let Dew go to her work. You will need a careful helper then, for the earth is thirsty and the plants need moisture to grow. Set her to work in the evenings, when she can take as long as she likes, and in the morning, all will admire her handiwork."

"And they won't mind working only for one season at a time?" asked Sister Rain.

"On the contrary," replied Mother Nature, "they will be grateful to be able to help by doing what they do best."

"Thank you, Mother Nature!" cried Sister Rain. "I knew I could count on you to help me." Mother Nature smiled pleased to know that she had helped her children and grandchildren once again.

VIII.
Composing Pedagogical Stories

Pedagogical stories require a commitment to meditate and reflect on our students. By their very nature, these stories need to speak directly to the student(s) for whom they are composed. The images and situations in these stories need to ring true; otherwise they will not touch the students' souls.

There are two types of pedagogical stories: those that address difficulties and those that offer hope or encouragement. If children have character traits or are behaving in a way that is not serving them, pedagogical stories can help them to see themselves and the consequences of their behavior more clearly.

Pedagogical stories can also offer hope or encouragement for a child who is struggling or who is facing difficult situations. Such stories help the entire class by strengthening classmates' compassion for the child facing the challenges and by helping everyone see the positive results of facing life challenges.

I have found the ideas expressed in Erich Gabert's *Punishment in Self Education and the Education of Children* extremely useful in addressing children's problems through pedagogical stories. Gabert's suggestions also helped me find positive—rather than critical or judgmental—approaches to the issues I wished to address.

Gabert states that to be most effective, our work on problems and our approach to punishment with children (and ourselves) should mirror the work of the angels. The angels' task is to help the human beings entrusted to them to pursue their destiny by realizing their higher self. They accomplish this task largely during the night by reminding us of who we really are and where in life we are heading. The common experience of waking up with contrition for what we did yesterday can be attributed to our meeting with our angel in the night. This reflects, in microcosm, what we will do after death when the angels and other beings help us form resolves for our next incarnation.

According to Gabert, parents and teachers can support the angels by helping children strive for their highest selves. We do this not by admonishment or punishment, but by holding up a loving mirror in which the child can perceive herself truly. If a child is allowed to perceive herself and the results of her actions, she will experience contrition, and a desire to make restitution will awaken in her.

Pedagogical stories are one way to hold up a loving mirror in which children can see themselves and their actions reflected. The stories provide children opportunity for self-reflection and suggest ways to approach a situation.

How do I go about composing a pedagogical story? First I try to develop a more objective view of the child(ren) or situation that needs to be addressed. Once I develop this objective view, I summarize the situation for myself. This will provide the theme for the story. Here is one of example of such a summary: Sally doesn't think anyone notices her or recognizes her fine qualities.

Next I begin to look for natural characters or situations that mirror the child's situation. You will be surprised how readily the natural world provides analogies (or parables) for our human problems and situations. And so many aspects of nature can offer hope—the daily rising of the sun, the turn of the seasons, the calm after a storm, the determination of a seedling to push through the pavement. If we attune ourselves to the world around us with a questioning soul, we will find the answers we are seeking.

Interestingly, when a situation or problem needs to be addressed quickly, the inspiration for the story often comes immediately, but certainly within a day or two. When a situation has been building for a while, or the story seeks to address an aspect of the child's character, the gestation period may be longer.

Once I have found the natural character or situation, I summarize it. In Sally's case, I found an example in the unrecognized worth of a single tree in the world of trees: A little fir tree sees all the other trees that are so colorful in the fall. But when their leaves fall, she realizes that she is the only tree who stays green all Winter.

With the theme and the character clearly expressed, I develop a plot and allow the characters to speak for themselves. (Sometimes they take the story in an unexpected direction.) For Sally's story I evoke sympathy for the little fir tree by describing how sheepish she feels about her drab dress while the other trees revel in their glory all Autumn. Then I tell about those beautiful leaves withering and drifting to the ground. I describe a stormy night and how in the morning, the little fir tree looks around and sees that all the other trees' branches are bare but she still has her bright green needles. I conclude the story by telling how proud the little fir tree that she stays green all Winter long and how the next Spring, when the other trees begin to leaf, she no longer feels sheepish and drab about her appearance.

Here is another example of a summary: James' life has become harder—his parents are getting divorced; his grandfather just died; and he is going to have to move.

I again found an example in the world of trees—the difference between a stand of pine trees that grows in a valley and a single oak that grows on the mountainside. A storm can knock down the pine trees in the valley because they are shallowly rooted, but the oak tree growing on the mountainside will withstand the storm because he grips the rocky ground with his strong roots.

Then I create the story: I describe a pine tree among his brother pines in the gentle valley. The pine tree has an uneventful life and grows quickly, not needing to put down deep roots because the valley floor is so fertile. How proud the pine tree is to stand among his fellows, so tall and straight, stretching his limbs toward the sun and the sky!

I then describe the life of an oak tree that springs from an acorn that a squirrel has buried on a mountain ledge. The oak tree grows slowly, buffeted by the strong mountain winds and enduring the harsh Winter snows. Each year he becomes more gnarled and bent, but his strong roots reach deep and grip the rocky ground with determination.

Once I have recounted these parallel lives, I tell about a storm that sweeps through the valley, knocking down the pine trees that cannot withstand the powerful wind. The storm batters the mountainside, but the little oak holds on.

Such stories do not need commentary—their message is self-evident. In both these cases, however, I decided to compose birthday verses for Sally and James. Because confidence and hope develop slowly, I thought that the reciting of the birthday verse for a year would reinforce the story's reassuring message.

I encourage you to try composing pedagogical stories for your students. The stories need not be works of art; they just need to ring true. You do not have to write them out (though I do because I gain confidence when I practice with a script), and the story can be more informal than some of the other ones you tell.

If you don't think you can compose your own story, seek about for examples that you can adapt and make your own. Be assured: You are likely to find what you are seeking.

A note about pedagogical stories after third grade: In the middle grades, you do not need to use a story form but can speak to the class about the ways you and other people have discovered how to face life challenges. Children older than nine find autobiographical anecdotes and stories of friends very interesting because they want to know their teacher better. In the upper elementary grades, biographies and biographical vignettes can serve as pedagogical stories by providing living pictures of how women and men met their life challenges.

Here is an example of a pedagogical story written for a third grade student. This student was not outwardly defiant, but he did not follow directions, did not come when called, and did not get along readily with his classmates. Reminders did not work, so I thought that it would be good for him to see the consequences of this attitude and behavior played out in nature. I chose geese as my characters because the young geese have to follow their parents' example and to work hard to become strong enough to go on a long, arduous journey twice a year. I briefly debated whether to have the story end with Bright Eyes' death (serves him right!), but my better self prevailed: Redemption is always better than punishment.

Bright Eyes' Journey

Far north, in the wilds of Canada, there lived a mother and father goose and five goslings. Both the mother and father were large and strong, and four of their children took after them. The youngest gosling was small, but he made up for what he lacked in size with energy and adventurousness. That gosling's name was Bright Eyes, and there was hardly a moment in the day when he wasn't up to something—flying, swimming, hunting for food, or, his favorite activity: exploring.

One day Bright Eyes found a shiny spoon left by a camper, and he carried it in his bill out to the island where his family nested. He liked the way the sunshine reflected off the bright metal, and he decided that he would start a collection of treasures. Every day he went looking to see if he could find one, and often he did.

Bright Eyes had one habit that worried his mother and father: He didn't always come when he was called. Geese always post a sentry to watch out for danger, and when the sentry sounds the danger call, all the geese quickly fly or swim to safety. Bright Eyes would sometimes come when he was called, but if he was in the middle of something interesting, he wouldn't.

One day a bear came out of the woods and started gathering blackberries from the bushes by the shore. Bright Eyes' father called, but Bright Eyes didn't fly right to him. Then it was too late. The bear was blocking his way, and there was no way he could make it to safety. Bright Eyes crouched very low among the weeds and hoped that the bear would not see him. Luckily, the bear was only thinking about stuffing his mouth with berries, and he didn't even notice Bright Eyes.

One day in early September, Bright Eyes noticed something shiny partly buried among some weeds by the bank of the pond. As he was pulling away the weeds, his mother gave the danger call. *It will just take me a moment to get this*, he thought, and continued to pull. All of a sudden, he felt himself knocked to the muddy ground. A fox had him by the wing and had started dragging him away. Bright Eyes struggled to break loose and honked and squawked, but the fox held on tight, his teeth biting into Bright Eyes' shoulder. Bright Eyes despaired that he would ever get free. All of a sudden, the fox was hit by a huge blow from the side. Bright Eyes' mother had hit the fox again with her mighty wing, and the fox let go. Then Bright Eyes' father arrived and started pecking at the fox's face. Realizing that he could never win a fight against two full grown geese, the fox let go and scampered off, disappointed.

Bright Eyes' mother and father stood by his side while he recovered. Then they started to fly back to the island where their other goslings were waiting. Bright Eyes didn't go along because he couldn't flap his hurt wing well enough to fly. Bright Eyes' mother circled back to him and swam by his side as he paddled out to the island in the center of the pond.

It was late September, and every day geese from surrounding lakes and ponds began to gather at the pond for their journey south. Bright Eyes' father and mother wondered, *Would his wing be strong enough to make the journey? What if it wasn't?*

The day came when the flock had to leave if it was to avoid the coming frost. Because Bright Eyes' father was the leader of the flock, he had to leave, but the rest of the family stayed behind with Bright Eyes to wait for another departing flock. His father told him to try to exercise his wing as much as he could so that he would be strong enough to join another flock on their journey.

Early the next morning, Bright Eyes' father and mother said their sad good-byes, and the flock flew off. Bright Eyes watched them go and then returned to his exploring. He didn't need to fly to hunt for treasures; there would be time enough to exercise later.

During the next few weeks, Bright Eyes didn't exercise his wing, in spite of his mother's reminders. "I'll exercise later," he said. "And besides, my wing will get better by itself." And slowly his wing did heal, but not quickly enough. He wasn't able to fly for more than a few minutes at a time, and every day the weather was growing colder. One morning, when the geese awoke, they saw that the edges of the pond were rimmed with ice. Soon the killing frosts would come.

The next day, a flock of geese flew over the pond. When they saw Bright Eyes' family, they landed to see who they were. After they had listened to Bright Eyes' mother's story, they said, "Join us! You need to leave soon, and you'll be safer with all of us."

Bright Eyes' mother anguished over what she should do. If they stayed in Canada much longer, she and her children would freeze to death. If they left now, Bright Eyes would have to stay behind. "I can't risk my whole family," she finally decided with regret. "I have to save my other children." The next morning, Bright Eyes' mother, brothers and sisters said good-bye. "I hope you'll be able to join us soon!" they said. Bright Eyes watched as his family flew off.

Now that he was entirely on his own, Bright Eyes finally realized how serious his situation was. *If I'm not strong enough to fly south, I will die*, he thought. All that day he exercised his wing as much as he could bear. The next day he exercised again, taking short flights until his hurt wing gave out. For the rest of that week, he exercised constantly, growing a little stronger and more confident every day. After each flight, he returned to the pond and stayed on the island for safety. There was no one now to warn him of danger. Every night, as he tucked his head under his wing, he thought of his family.

The days grew colder, and Bright Eyes knew that he would have to leave before everything froze. He didn't know the exact way, but he knew the general direction. Each day he flew southward as far as he could. Each day, his wing grew stronger, and he was able to fly a greater distance. Each night he found a safe place to sleep. One evening, as Bright Eyes was resting on the banks of a river island, he heard a distant honking. A flock of geese was coming in for a landing.

"Where are you coming from?" he asked.

"Northern Canada," the leader replied. "We were delayed because of the drought. There was so little food that we were too weak to fly very far each day."

"Can I go with you?" Bright Eyes asked.

"Of course!" the leader answered. Bright Eyes was overjoyed to be part of a flock once more. The next morning he flew off with them.

Bright Eyes was no longer alone, but he was lonely for his family. Day after day, as he flew south, he wondered: *Will I ever see my family again?* When the flock finally reached their Winter feeding grounds on the Chesapeake Bay in mid-December, Bright Eyes thanked them and took his leave. He was going to find his family.

Each day Bright Eyes flew over the bay, looking for them. Lonely and disappointed, each night he found a safe place to sleep. Late one night at the end of December, he was awakened by the peals of church bells. *What are they celebrating?* he wondered. *Maybe that is a sign that I will find my family.*

The next morning, the sun rose bright and clear. Hoping this would be the day he would find his family, Bright Eyes flew off,. He flew all day, stopping only once to eat. Late in the afternoon, he spotted a group of six geese feeding among the rushes. He swooped down low to look more closely. It was his family! With a loud honk, he landed among them. For a moment, Bright Eyes' mother and father didn't recognize their son. How large and strong he had grown! Bright Eyes' brothers and sisters crowded around him, showering him with questions. But Bright Eyes couldn't answer; he was too full of joy to be back with those he loved.

IX.
Working with Birthday Verses

Background

Birthday or report verses hearken back to indications given by Rudolf Steiner at a faculty meeting in 1921. When speaking about end of year reports, he said, "At the bottom of each report, write a verse for each child that expresses the child's individuality, that can act as a leitmotif for the future." (*Faculty Meetings with Rudolf Steiner,* 252)

Heinz Müller worked with Rudolf Steiner to bring this indication to fruition. In *Healing Forces in the Word and Its Rhythms,* he describes how to use imagery, wording, sound, and meter to enhance the power of the verse.

I was so daunted by what Heinz Müller described that I almost gave up on the idea of writing report verses. I couldn't imagine writing dozens of verses (in addition to reports), crafting each one so carefully so that it would give the child a sense that he or she was truly known.

It wasn't clear to me whether report verses were meant to be read or heard just when the report was received, or whether the child would recite or hear that verse more often. I could imagine it becoming part of the child's bedtime routine, perhaps accompanying an evening prayer. Such repetition would strengthen the power of the verse. But I didn't know whether parents would follow through, whether children had a bedtime routine and if so, whether a verse would fit into that routine. So I decided to give each child the verse on his or her birthday, although I included it in the year-end report as well so that years later, a child would encounter the verse again. This would accomplish several goals: It would spread out the task throughout the year, making it more manageable to compose many verses, and it would make the child's birthday more special.

I also decided that because the power of a verse is intensified by repetition, students would recite their birthday verses on the day of the week they were born. They would do this throughout the school year until their next birthday, when they would receive a new verse. (We would celebrate Summer birthdays as half-birthdays, and students who are born on Saturday or Sunday would say their verses on a chosen day.) I hoped that the impact of the verse would be magnified by having the other children witness the recitations and unconsciously perceive how the verse reflected the soul of the child.

I imagined that first graders would be capable of learning their verses and reciting them alone easily, but the child who had the first birthday of the year—a shy girl—needed me to recite the verse with her. For a few weeks I recited her verse with her. After the third time, the rest of the class chimed in and said the verse too. By the end of the year, each day of the week had its special character, enhanced by the class recitations of the verses of the children born on that day. This continued until the child's third-grade birthday. When I presented the birthday verse, I announced that this year the child would be reciting her verse on her own. We therefore slowly transitioned to a process where each child had a special moment in the week when she would say the verse alone while everyone else listened.

With my second class, students wanted to recite the verse with the birthday child long after third grade, so I didn't insist on individual recitation. Some children announced that they wanted to say their verse alone; others continued to want the class to say the verse with them for years. Another practice that began with my second class was that when a child was absent, we all stood up and recited the verse in her honor. That brought the absent child more fully into the class's consciousness.

One further detail: In order to make the birthday verses more special for the student and the class, I created illustrated versions which were compiled into a "Birthday Book." First and second graders loved to look through the birthday book, and several of them learned to read by reviewing the many verses that they already knew by heart. It should not surprise you that even in sixth, seventh, and eighth grade, students will enjoy leafing through the birthday book and recalling the verses that they and their classmates recited over the years. And a packet of a child's collected verses is a meaningful present when the child graduates or leaves the school.

After trying to apply Heinz Müller's suggestions of how to work with all elements of poetry to make the verses more potent, I realized that I didn't have the understanding or the skill to do so. Instead, I confined myself to finding an image or a parable that I felt the mirrored or expressed the child and offer the child soul sustenance for her journey.

Composing a birthday verse that is "true" requires that we reflect and meditate on our students and that we allow sufficient time for inspiration to come. Over the years I found that if I began seeking for an image that would speak to the student at least two weeks before the birthday, I was likely to find it.

The process of composition also benefited from extra time because words, phrases, and rhymes had a way of suggesting themselves, often when I least expected them. On one occasion, I had been agonizing for several weeks over a somewhat self-centered first grader's verse, seeking an image that would awaken greater compassion for her classmates. One afternoon while I was driving from Richmond to Charlottesville, I suddenly heard her verse echoing in my soul. I pulled over and wrote down the following:

> The seeds of compassion lie buried
> Deep in the human soul;
> In the light of love they flourish and grow
> Into deeds of grace untold.

Another time, I had been agonizing for weeks to write a verse for a child who had significant developmental challenges, but who was the light of our class. A few days before his birthday, I woke up in the morning with this verse resounding within me:

> The Light of the World came down to earth
> To shine for all Mankind;
> I carry that light within myself,
> In my will, in my heart, and my mind.

Over the years, a few verses came readily to me, but most didn't. As I became more experienced with the process of composing them, I realized that I needed to trust that a verse would come; I just had to give it time. Although it was frustrating to wait for the verse to ripen, I consoled myself with the knowledge that my reflection and meditation on the child would reap benefits for the child and for me. And though I was sometimes dissatisfied with the final result, I knew that the child on whose behalf I was working would benefit from my striving to offer something of value.

I know that many teachers will think that they do not have the talent or skill to write verses for their students. Or they may think that since they didn't start the tradition in first grade, it's too late to start later. Both are not true. One can start working with birthday verses any year (though sixth-, seventh- and eighth-grade students are more responsive to poems and prose passages by other authors). And every teacher is capable of creating heartfelt messages for her students. I strongly encourage you to try; you may be surprised by what you are able to create, and you will definitely be gratified by the students' responses. And in those desperate moments when you don't feel up to the task, you can take solace in Ben Johnson's maxim: "What is written without effort is read without pleasure."

Although birthday verses composed by a teacher may have special meaning for her students, verses from any source can be equally meaningful and effective. Heinz Müller includes many examples in his book, and *A Journey Through Time in Verse and Rhyme* by Heather Thomas also offers many verses. I have included verses in Appendix 3 to give plenty of examples for emulation and adaptation, or simply used as written.

Composing your own verses – a primer

Although I don't consider myself an expert in composing birthday verses, I certainly have had plenty of experience and done my share of agonizing. In this section I will try to describe how I approached this challenge in the hope that it will support your creative process.

In my opinion, the most important aspect of the birthday verse is the central image. This image should express who the child is or is becoming. When children receive verses with the right image, they feel seen and known, and they are reassured that we understand and love them.

Lest you think that we can find images only through some mysterious, esoteric process, let me reassure you that images are all around us—we just have to begin to pay attention in a different way. Every natural phenomenon can speak to us and reveal its message if we but listen with our souls. The same types of images that we used for nature stories, pedagogical stories, and parables can serve in birthday verses.

Here is an example of a situation that I wanted to address: A child's family was going through a divorce, and she was feeling deeply unsettled. When I looked for an image, I thought of a ship helplessly tossing in a tempest. The image of the tempest-tossed ship led me to think of the ship's anchor. This child needed to be reassured that she was well-anchored.

With these thoughts in mind, I sketched out my idea: A ship that is being tossed about by the blasts of a storm needs an anchor to hold it. What anchors us in our lives? Our relationships with one another. I tried to express the image poetically. Finally I arrived at a statement of the image: "Firm the anchor holds the ship." Then I asked myself: What does the anchor do? It keeps the ship secure. It stops the ship from drifting away. I tried some synonyms and came up with *sure* and *fast*. The first part of the verse was now "Firm the anchor holds the ship and keeps it sure and fast."

Then I tried to express the correspondence between *relationships* and *anchor*. I cast about for the best word for *relationships* and decided to try *bonds*. This matches the idea of an anchor on a chain. The expression, "So the bonds that link us all," came to mind, and I held it as a possibility while I tried out various possibilities for the last line.

The necessity of rhyme often works magically to help us find the right words. To rhyme with *fast* I chose *blast* to express the storms of life. Now I had a final challenge: to express what the bonds that link us all allow us to do with respect to the storms of life. After trying several expressions, I settled on the verb *weather*, pleased with its double meaning to the storm and to making it through the storm safely. The second part of the verse now became "So the bonds that link us all will weather every blast." I agonized whether I wanted to use the verb *will* or *can*. *Will* had the advantage of alliteration, but this child was a melancholic and if the verse were too affirmative, I thought that it might not ring true to her. So I settled on the verb *can* which leaves a little doubt. I hoped that the reassurance of the verse would come from repetition.

The final version read:

> Firm the anchor holds the ship
> And keeps it sure and fast;
> So the bonds that link us all
> Can weather every blast.

In eighth grade, when we reviewed our years together as a class and discussed the experience of reciting and hearing birthday verses, this child told the class that this verse had sustained her through the years, and she predicted that she would never forget it.

Birthday verses can also express hopes or aspirations. The process of finding an image for such verses is identical to the one described above. Once you have found the image, you need to articulate the hope or intention that the image inspires. This step benefits from making a prose statement before trying to put it into the verse. For instance, I decided that a good image for a timid girl was the moon, which starts off as a crescent but gradually gains in brightness. The hope to be expressed was: I will try to be brave enough to show the world who I am.

After some poetic wrestling, I came up with this verse:

> The crescent moon so meek and mild
> Prefers to hide her face;
> But when she waxes full and bright
> She shines with strength and grace.
> O may I too be brave enough
> To shine with all my might;
> And like the lovely silver moon
> Fill all the dark with light.

If my process of composing verses does not reflect yours, I will not be surprised. Each of us has a different way to tread the path of creation. Whatever your process, be patient with yourself and take joy in your efforts. Any verses you create will be received by your students with gratitude and will nurture their souls.

X.
Practical Aspects of Working with Home Surroundings

Main Lesson Blocks

In the first Waldorf school, there were only three subjects taught in the first three grades: Language Arts, Mathematics, and Home Surroundings. According to Karl Stockmeyer's *Curriculum for Steiner-Waldorf Schools*, Language Arts was to be taught for 14 weeks, Mathematics for 12, and Home Surroundings for ten. Adapted to our shorter school year, we should devote about eight weeks to Home Surroundings.

I recommend having two Home Surroundings blocks in first grade and two in second grade, ideally one in the Fall and one in the Spring of each year. During those seasons, the natural world goes through the greatest outer changes, and because those seasons serve as the balance points of the year, we can always refer to the previous season or look forward to the next.

Having taught main lesson blocks of all different lengths, I have concluded that four-week blocks are best whenever possible, especially for subjects such as Home Surroundings, Physical Science, Geography, and History. The month is an ideal rhythm for children's learning; anything shorter feels abrupt, and the block content won't be as fully digested nor live as strongly in the children's memories. Blocks longer than four weeks work in first grade, but after a month, second graders might become impatient to move on to the next subject.

Home Surroundings Block Content

For first grade, I recommend telling a mixture of descriptive stories and stories with a moral element about the heavens and the heavenly bodies, day and night, the seasons, the weather, and the landscape and streams, rivers, lakes, and the ocean. Stories about plants and animals through the seasons can be interspersed to give more flavor.

For second grade, I recommend descriptive stories that are more down to earth and deal with cycles of nature and the characteristics of different plants and animals. Origin stories work especially well in second grade because they are often humorous. Second graders are still receptive to stories with a moral element as long as they are not "babyish" in any way. Fables and animal tales can be included in the second grade Home Surroundings curriculum, leaving the legends of virtuous people for the Language Arts main lessons. In my experience, devoting one block to nature stories and the other block to fables and animal tales works well. Fables and animal tales can also be told at the end of the lesson during Math blocks.

Types of Presentations

In first and second grade, we can present stories in two ways:

    a) told with a brief introduction or just the title
    b) preceded by a conversation that illuminates the themes and characters

Stories in the first category engage children on a more unconscious level; the children dream their way through the story. Stories in the second category call on children to be fully conscious because they are being asked to respond to the teacher's comments and questions during the introduction.

I think that descriptive stories, stories with a moral element, and fairy tales and legends should be received with a dreamy consciousness and should stir the depths of the children's souls. Fables, animal tales, and origin stories benefit from being previewed and discussed to engage the students more actively and consciously. Suggestions for how to conduct such conversations are given in the chapter on fables.

Reviews

The type of review we conduct depends on the type of story we have told. For a fairy tale or legend, we want the children to be able to account for the plot and to describe the characters. The setting of the story often supports the action but is not of primary importance. Especially in first grade, I told fairy tales, legends, and nature stories with a moral element more than once and as many as three times. Repeated hearing deepens the child's experience of the story and leads to reviews that are more profound and meaningful.

Depending on the students and the class, reviews can lead to discussions of the theme and expressions of opinion. Nature stories with a moral element and some descriptive stories benefit from this kind of review.

When we review animal tales fables, and origin stories, the plot is only important insofar as it helps the students understand the characters and the themes. I usually told these types of stories only once, because their plots were simpler and their themes less profound.

Such stories sparked lively reviews, and our conversations often digressed into uncharted territory. Questions about the story abounded, and children were asked about their experiences. The conversation before the story and the review served as bookends to the story, giving the students many opportunities to relate to it.

Reviews can also include "free renderings" or drama. When stories are dramatized in the review, whether entirely with a narrator or partly with dialogue, they take on new life, and students relate to the characters on a different level.

Main Lesson Book Work

The Home Surroundings study lends itself to highly illustrated main lesson book work. Especially in first grade, when students are not yet able to write very much at a time, a drawing and a caption from a story will suffice. I think that nature drawings in the early grades should not strive for realism; rather, they should express the gesture and the feeling of the subject.

In second grade, there are several more options for main lesson book work. Students can copy from the board or take dictation, but Rudolf Steiner stated in the Curriculum Lectures that they should begin to write their own short accounts of what they have heard. Their short accounts can imitate the teacher's, and the teacher can create a model of economy by writing pithy captions or short poems about the subject.

Children can copy or take dictation of a short poem or verse. For example:

> How gracefully the great geese fly,
> Fly across the sky.
> I love to hear their joyous cry,
> Sounding from on high.

Poems need not rhyme or scan; in fact, cultivating poetic speech and a freer approach to versification will serve the children well throughout their education. When they try too hard to rhyme, it is easy for children to write hackneyed verses. Instead, I recommend that children be inspired to write poems that are pictorial and evocative, such as:

> Graceful geese, flying south,
> To their safe, warm, Winter home.

Fables, animal tales, and origin stories lend themselves especially well to short written pieces. For instance, for the fable "The Fox and the Crow," children could accompany a picture of the fox and the crow with the caption: "Foxes are clever; crows are proud and gullible." Or for the Russian fable about the scorpion and the frog, they might write:

" 'Why did you sting me?' asked the frog. 'It's my nature,' said the scorpion."

Some teachers have their students keep a nature journal to chronicle nature walks and seasonal changes. Such a journal is an excellent opportunity for students to write their own accounts accompanied by drawings or to draw what they have experienced accompanied by captions or brief remarks. I do not believe that all the children's work needs to be correct or corrected—it's important for them to become comfortable "talking on paper"—and too much focus on spelling, grammar, or handwriting can stifle a child's writing. In my opinion, what is copied from the board or taken as dictation should be accurate (and corrected if necessary), but children's original writing can remain an uncorrected original creation.

Morning Exercises

We should include some nature poems, songs with a nature theme, and appropriate movement exercises throughout the year as a way of helping our students connect with the seasons. During Home Surroundings blocks, these morning exercises can be focused on what we are studying, thereby enhancing the story content.

There are many sources for poems about all aspects of nature, and teachers should consider carefully which are suitable for first and second grades. My primary criteria are vivid imagery, beautiful language, and the ring of truth. I highly recommend the anthologies compiled by Heather Thomas (*Journey through Time in Verse and Rhyme*) and by David Kennedy (*The Waldorf Book of Poetry* and *The Waldorf Book of Animal Poetry*). These collections have a wealth of appropriate poems on nature themes. And of course, if a teacher feels inspired to write original poems, these will be received by the children with gratitude and joy.

In terms of music, teachers will need to decide whether they want to restrict themselves to pentatonic songs or songs in the "mood of the fifth." If so, the following:

Collections of traditional children's songs that have nature or seasonal themes include *Sing through the Day* and *Sing through the Seasons*, as well as many others. See Appendix 4.

Teachers may be surprised to find how easy it is to set a nature poem to music to be sung or played on the flute. I encourage teachers to attempt to do this because, just like original stories, original songs and poems touch the children's souls deeply.

As for movement exercises, almost any verse or song can provide the foundation or accompaniment for movement exercises, dances, bean bag drills, or rhythm stick work. Engaging the movement teacher to have the children play games based on nature themes will further enrich our work with the children in the Home Surroundings.

Nature Walks

Nature walks are nourishing to our students, and I recommend doing them regularly, and, when possible, taking the same walks and revisiting the same places during different seasons. Whether your school is a city, suburb, or urban school, nature is all around if we but take the time to notice it. In fact, a city walk may help us become more attuned to the natural world because the buildings, roads, and sidewalks seem to suppress nature.

A nature walk is an opportunity for children to experience nature, not to learn facts about it. Even though you may be tempted to use the outdoors as a classroom, I urge you to simply join the children in noticing and marveling about what they encounter.

Rudolf Steiner was adamant about allowing children simply to notice and appreciate nature outside and confining the lessons about nature to the classroom. (See *Practical Advice to Teachers*, pp. 45–46.) Some children will notice what we have spoken about in class, and they will want to share their observations; others will have questions. Rachel Carson's *A Sense of Wonder* gives excellent advice about how to be in and engage with nature with children. Also George Russell's collection of essays, *Children and Nature*, provides ideas and inspiration.

Some children truly take in their surroundings during nature walks; others don't notice as much because they are busy socializing. When I would go on nature walks with my students, I would encourage them to focus, and sometimes we just sat still and looked around without talking. But I resisted the urge to use the surroundings as a way to capture their attention.

If you have the time to take the walk on your own prior to the class outing, you will notice much more about the surroundings than when you are also having to keep an eye on them. Then during the nature walk, you can draw their attention to what they might otherwise not notice.

If you have the time and inclination, try to attune yourself to the nature-spirits in the places you are visiting. This will deepen your experience of the place immeasurably. For indications on how to get to know the nature-spirits, refer to the article on "Nature Spirits" by Frits Julius in Appendix 2.

Nature Objects in the Classroom

There are three main ways to include natural objects in the classroom: nature decorations, the nature table, and displays of what the children have gathered.

Nature decorations enliven and beautify the classroom, and I encourage you to include natural treasures as part of the classroom environment. These treasures can change with the seasons or remain displayed all year.

I had a formal nature table in my classroom from first through third grades. This table was covered with a beautiful colored cloth that was changed every season to reflect the mood of the season. The nature table had a different theme every month, and sometimes treasures were added during the course of the month. A nature table can be graced with a beautiful card in a special holder or other objects that make the display more beautiful and more special and unique to the class of students.

Children will want to bring in objects they find on their walks and sometimes special objects from home. I think it's best to have a special place where these can be displayed rather than having them added to the nature table. When children bring treasures from home, make a special moment when they can tell their classmates about the object and where and how they found it.

Nature Crafts and Cooking

Nature crafts can enhance students' experience of their Home Surroundings. I suggest devoting several craft and cooking lessons each season to let the children make something that expresses the season. These lessons can be held during the second half of main lesson, a free period, or in lieu of a nature walk. Several books on nature crafts and recipes for cooking seasonal treats are included in Suggested Sources, Appendix 4.

Appendix 1
Rudolf Steiner's Home Surroundings Curriculum Indications
in the original German

from GA 295 *Erziehungskunst. Seminarbesprechungen und Lehrplanvorträge*

Nun wird man zu dem eben Gesagten dasjenige hinzufügen, was das Kind anregen kann zum Nachdenken, indem man ihm Naheliegendes erklärt: dasjenige, was später geordneter auftreten soll in Geographie, in Naturgeschichte. Das erklärt man ihm, bringt es meinem Verständnis nach, indem man an Bekanntes—an bekannte Tiere, an bekannte Pflanzen, an bekannte Bodenkonfigurationen, an Berg, Fluß, Wiese—anknüpft. Die Schule nennt das Heimatkunde. Aber es handelt sich darum, daß man gerade im allerersten Schuljahr ein gewisses Aufwecken des Kindes gegenüber der Umgebung zustande bringt; ein Aufwecken des Seelischen, so daß es lernt, sich selber wirklich zu verbinden mit der Umgebung.

Und wenn dann das zweite Schuljahr angeht, da wird man versuchen, das Vorerzählen, das Nacherzählenlassen fortzusetzen und weiter auszubilden. Das Kind kann allmählich im zweiten Schuljahr dazu übergeführt werden, daß es dasjenige aufschreibt, was man ihm erzählt. Und dann kann man es, nachdem es herangebildet ist an dem Aufschreiben dessen, was man ihm erzählt, auch veranlassen, das, was man ihm beigebracht hat über Tiere, Pflanzen, Wiese und Wald der Umgebung, in ganz kleinen Beschreibungen wiederzugeben.

Nun, Sie haben gesehen, wir verwenden in freier Weise dasjenige, was aus der nächsten Umgebung bekannt ist, um eben einen freien Sachunterricht zu treiben. Das Kind kann ganz gut, indem es mit dem dritten Schuljahr gegen das neunte Jahr zugeht, durch diesen Sachunterricht eine Anschauung davon haben, wie man—nun, ich kann nur Beispiele herausheben—Mörtel zubereitet, wie man ihn verwendet beim Hausbau. Es kann auch eine Vorstellung davon haben, wie man düngt, wie man ackert, wie der Roggen, der Weizen aussieht. Kurz, in freier Weise läßt man das Kind eindringen in dasjenige seiner nächsten Umgebung, was es verstehen kann.

Appendix 2
Nature Spirits
by Frits H. Julius

reprinted from *The Golden Blade*, 1971

To begin with, this article was intended for teachers, who have the task of introducing Nature to children through fairy stories. But it is intended as well for all who, because of their work or for inner reasons, are trying to deepen their connection with living things. It was a teacher, beginning with the first class in a Rudolf Steiner School, who urged me to undertake it. This teacher felt that it is not enough to talk to the children in a vague, general way about gnomes and elves; but that in this subject, as in others, a definite foundation of knowledge is needed. A good answer to such a question would not only be helpful to many teachers, but to mothers with young children, farmers, gardeners, and those who have to do with the care of the sick and the use of medicaments.

On this subject I cannot regard myself as a final authority—this is not offered as the work of a Master, hardly of an apprentice, but of a pupil; it will often therefore have too personal a character. It should be read with some reserve, choosing out what can be made fruitful. Good use can sometimes be made of quite personal things, if they are told honestly and sincerely. Everything described here goes back ultimately to Rudolf Steiner in two ways. On the one hand, I have collected, and briefly discussed, some of his most important accounts of this subject as a help to further study and as a general orientation. On the other, I shall describe some of the things I have experienced myself when following up his indications and the conclusions I have drawn.

It seems as if there is a renewed interest in this subject at the present time. And all such beings remotely of this kind appear in children's television, film, cartoons, and so on, more frequently—but in what I must call a trashy way. We should try to replace this with something better.

Anyone who wishes to go into this subject properly will find it necessary to lay a firm foundation through serious study. It is in three lectures from his course *Man as Symphony of the Creative Word* that Rudolf Steiner spoke in the greatest detail and most profoundly—so far as my knowledge goes—about the Elementals. It is important to notice the different points of view from which he describes these beings. Here he gives us a many sided and finely shaded picture. He begins by following the line of the plant's development through the four elements—earth, water, air, and fire. He lets us see how the nature-spirits live in these elements and are occupied with forces at work around the plant. He calls these beings (in the sequence of those elements with which they are particularly concerned) gnomes, undines, sylphs, and fire-spirits—rather than using for these last the traditional name "Salamanders.'...

Rudolf Steiner shows how the gnomes working on the plant handle the forces of the life-ether, undines use the chemical ether, sylphs the light-ether, and fire-spirits the warmth-ether. As well as their connection with the plants, their intimate relationships with animals are discussed in many different ways. Steiner goes on to the changes in these beings through the course of the year, their relationship to the earth as a whole, and their significance for the hierarchies. Then comes something very important; how these beings help in the formation and maintenance of our human organism and in what ways they are around us day and night.

The three lectures from *Man as Symphony of the Creative Word* should be considered within a whole sequence of lectures given at about this time (Autumn and early Winter 1923) in Dornach, including *The Four Seasons and the Archangels* and *Mystery Knowledge and Mystery Centres*. It is very important too that Rudolf Steiner also gives us some mantric sayings, through which we can deepen our understanding of this subject in meditation. Some of these sayings express what these beings feel about themselves, some are questions or challenges to awaken, directed toward us.

If one tries, year after year, to deepen one's understanding of these lectures, approaching them in meditation and with intensive imaginative activity, from many points of view, one can begin to build up an impressive picture of the life in their realm. This is very necessary for orientation. For their realm is a bewildering one; it has to be remembered that there are very many kinds, many categories, among nature-spirits, making them as a whole difficult to survey.

Apart from study, how can we achieve a personal connection with these beings? This is not the same as learning to perceive them in a true imaginative consciousness. It is quite possible to have a fruitful exchange with these beings without any, or hardly any, actual "seeing" of them. Before considering some indications from Rudolf Steiner about this, I would like to touch on an instructive personal experience. It happened more than thirty years ago, and its objective significance has been tested.

After I had studied biology, I felt for some time that I should find some quite fresh way, of a more inward kind, of approaching the metamorphoses that take place in plant growth—a way that would lead more deeply into life itself. At the same time a person of considerable experience was advised by a doctor, as part of the treatment of an illness, to practice metamorphosis in thought. We decided to give each other support by working at this together. Each day we took before our imagination a plant on which we had agreed, and passed through the whole cycle of its development—a snow-drop, for example, a wild rose, or a poppy. Of course we took plants we knew well, or those about which it was not difficult to discover missing details from nature or from pictures and descriptions. These exercises produced quite soon a surprising impression—I met within me, but coming from nature, a wave of joy, like a festival.

One Summer day we chose the birch-tree. It began as a small, bare tree. Then we let the buds unfold. No sooner were the leaves fully outspread, hanging on their fine stems than a mild breeze blew through the tree and set them all in fluttering movement. This came as a surprise which was increased when I learned later that the other person had had the same experience. I felt that nature itself was teaching us. I realized that on this path, one could be led to surprisingly essential things and toward a real goal. Something like a key was being offered to me through which countless treasures could be unlocked.

I will sum up some of the lessons which followed: One should never look at things as separate; they can only be truly grasped in their whole interaction with the environment. The birch for instance has a special connection with the airy element. And in general one should watch the play of the wind with leaves and twigs, catkins and sometimes flowers, whole trees and shrubs. Remarkable impressions of the different characters of plants can be gained in this way. But one can also learn quite clearly how playful the nature-spirits are. I can recommend everyone to try this. It can very much enrich one's picture of nature.

The light movement of leaves belongs particularly to the Summer. But one can look for the tree's response to the four elements at other times of the year as well: Autumn reveals in particular in the changed colors the effect of warmth. In Winter one sees the withdrawal of the tree into earthly existence alone, with the sharply outlined construction of its form. In Spring comes the tender work of the watery element in and around the shooting buds. Of course, to reach all these conclusions from the starting point indicated here requires knowledge of a certain kind and intuition—and a real will to investigate such things.

This was the beginning of what I would like to call a "gnome university." All that we had learned at the human university acquired new life; the acquired new life came into harmony with the order of nature and was raised to a new level. Nature began to shine for us, to regain the splendor we felt in childhood and youth, which had largely vanished during the years of study at the human university.

What I have described runs parallel with what Rudolf Steiner indicates in *Knowledge of the Higher Worlds* as the forming of "right conceptions." The first step was done by us when we brought ideas into movement in the right way. Then nature confirmed the value of this activity, coming to meet us, and even correcting what we were doing by taking a concept out of its isolation.

When the inner activity of forming concepts is brought into accord with the real happenings in the world, the veils which hide from us the realm of creative forces become thinner and gradually transparent. Thus one is impelled to think afresh about the minerals and the animals as well, that their kingdoms may become transparent too. And on a direct line from mobile conceptions of plants, one can come to the picturing of planetary movements, with far-reaching results.

Even such simple exercises have certain dangers, as I noticed later on. As one's conceptions acquire greater reality, they have an increasing effect on one's own organism. This can be very favorable; but it can be unfavorable too. When one is concerned with the growth of a plant, one may easily put too much emphasis on the processes of unfolding, of all that extends into space. Through this a kind of giddiness can be felt, as if one were being lifted out of the body. A balance is brought about by a greater attention to the process of contraction, where the plant is gathered together into the seed, along with the material densification and hardening that belong to this.

An indication which can be used by anyone, and which also brings a rich harvest, if it is really followed up, is contained in Rudolf Steiner's lectures on *Spiritual Beings in the Heavenly Bodies and the Kingdoms of Nature*. He says that impressions of the beings behind nature can be obtained if one develops a "moral feeling" for one's environment. He explains what he means by this through very illuminating examples. He also gives indications specially directed toward different groups of nature beings.

On a cloudless day, one can give oneself entirely to the impression received from the blue sky, excluding all other perceptions, memories, and so on. If one does this with sufficient intensity, a moment comes when the color-impression itself disappears and is replaced by a sense of reverent, unending devotion. In this way, a sense impression becomes a moral feeling. After providing some other examples, Rudolf Steiner says that in order to learn about the gnomes, one should go down in a mine (especially where metals are worked) and concentrate on the impressions to be found there, in just the same way as with the blue sky—and particularly upon the fine differences of warmth that can be felt.

An experience of the undines can be approached in the same way if, in very changeable, windy and misty weather, one observes the effects of moisture in the atmosphere. Or one can watch the spray of waterfalls.... Sylphs have a special task where living things ripen and decay. To have a clear impression of them, one should concentrate on air that is not moist, but clear, drenched in sunlight. Fire-spirits are to be met in general where life develops through germination.

If the inner attitude and activity of "moral feeling" have been practiced, keeping closely as befits beginners to the directions and examples given, a wide range of possibilities opens up. One can immerse oneself, for instance, with utter devotion in the color of a flower, small though it is in relation to the sky. All natural phenomena should be faced in such a mood as far as one can achieve it, or at least with an attentiveness and reverence which lead in this direction. Then nature is given the opportunity to speak to one. One should not reject anything of its revelation through over-seriousness, or out of too theoretical or traditional thoughts; one's whole activity should be transformed into a devoted receiving.

To complete this process, Rudolf Steiner gives as an example concentration upon a musical interval. In one's picture of nature the sounds must, of course, be included: the fir tree's murmur, the rustling of the leaves, the brook's conversation, the call of the animals. In *Knowledge of the Higher Worlds,* he provides very specific indications about listening to sounds, at first those made by animals.

In a Whitsun lecture given at Cologne on June 7, 1908, Rudolf Steiner directs our attention to places where two realms meet. Nature-beings seek such places, and appear there most clearly. Gnomes for instance can be observed where a vein of metal meets stone, undines where water trickles over a rock. Sylphs can be sought at the moment when a bee sucks nectar out of a blossom. Fire-spirits appear especially when man has an intimate connection with animals—a shepherd, for instance, with his sheep.

Such indications are valuable, quite apart from the development of clairvoyance. We will get to know nature more intimately if we practice a sort of encounter with the nature-spirits, if we look out everywhere for such frontiers, such borderlines. In the wind that moves the leaves we already have had a beautiful example of this. For people living by a shore, there are the waves which run up the strand and then flow back again. Looking at the foaming breakers, the play of nature-spirits can be discovered; very strong impressions are possible of their extraordinary, extravagant delight.

There are people who prefer to go out alone, seeking their most intimate experience of nature. Others receive their strongest impressions when they wander with someone else, if there is harmony between them. If there is a bad mood, or even indifference between them, it is as if a veil is covering everything. If the mood lightens, nature can open up again and sometimes even arrange special festivals for us. This happens most of all in the company of a young child. Over many years, such great moments have come even on excursions with groups—but only when the whole group is so mutually harmonious that it becomes like a single being.

In the third lecture of the course on *Mystery Knowledge and Mystery Centres* Rudolf Steiner explains how "dialogues" with nature-spirits can be achieved. When this is really attained, it has special importance for education: The power of fantasy is awakened, which can inspire the making of new fairy tales. Of course, something has to be said to the nature-beings that will interest them; otherwise there can be no conversation. Now the gnomes and the undines are specially interested in children under seven years old. After this age, the human being brings into his sleep at night something of himself which enters the world of these beings. The child under seven does not do this yet. For gnomes and undines the young child still belongs to "the beyond." They are therefore mightily interested if one tells them something about young children. Through this, one does them a great favor. (For details see the lecture itself.)

I knew that I ought to follow this indication, but I never got down to doing it. I've often heard of people trying to achieve this contact with nature-spirits for other than educational reasons—for instance, when they were searching for crystals—and having good results. Then years ago I learned of a small gathering of young people high up in the Lötschental in southwest Switzerland under the leadership of a friend. Rudolf Steiner's first mystery play was being studied. Quite early in this play comes the scene where the Spirit of the Elements demands a fairy story from Frau Balde. In the group discussion of this scene, much came up that was connected with the telling of stories to children.

Soon afterward, an expedition was planned up into the region of snow and glaciers. I was very much looking forward to this. But most unfortunately I had to retire to bed with whipped cream poisoning. (Even the best cream can act as a poison if not taken homeopathically.) On the day of the expedition I barely able to struggle to my feet. Exceedingly envious, I saw them go off, while I expected a boring and lonely day. But then I got a stroke of inspiration: They go up—I shall go down! Since I had no people to walk with, I thought, well then, I will choose the nature-spirits, and perhaps they would be better company. "If that fellow got enough contact with the nature-spirits for him to find crystals, they are sure to show me something too." In a few seconds all my rancor and envy were transformed into shining optimism.

Calling to my mind the best memory pictures I had of young children, I opened the door and stood on the steps, inviting the gnomes and undines as if I were calling children to a meal. And, O wonder, I walked then for three hours, without taking any step which was not led. Without actually seeing much of them, I had the whole time a very definite feeling, as if two gnome-like beings were taking me with them by the hand. I felt, through a constant gentle pressure or pull, impressions that were more moral than mechanical, what their intentions were. Everything depended on careful attentiveness on my part. However strongly and consistently I was guided, there was not the slightest compulsion. I had to think again and again how much we often lose through not being attentive enough.

First we went up, and then across a sunlit meadow, zig-zagging from flower to flower. It was already far into the Autumn, and I had not yet looked at any flower there. Thus to my great astonishment the nature-spirits did not show me crystals, but flowers. It seemed as if they had made a quick agreement: "This one here thinks he knows something about plants. We shall let him look for once at what he never sees." Then we went down to a stream between steep banks. I was put in front of a huge flat rock and made to look. In a great cleft in the rock stood a tiny tree. With the other surrounding small plants, it created the impression of a miniature garden. When I had expressed inwardly that I had seen and admired it, I was led further. This showed me clearly that these beings very much appreciate interest in their tasks and concerns.

In *Man as Symphony of the Creative Word*, Rudolf Steiner speaks of mediation between plant roots and the mineral realm as one of the main tasks of gnomes. Where can this be better seen than in such a place where rock is loosened and softened to become accessible for plants? Later on, they often put me in front of some nearly bare stone, on which only a few little plants were growing or where a patch of moss was beginning to claim its place. They pointed out where there was even just a breath of moss on walls along streets in our towns.

Another important task for gnomes is to awaken life in the seed and push the first shoots up into the light. So one can understand why later on they drew my attention not only to many stones but to bare places in the woods, where perhaps a single plant was beginning to grow. And even much later on, it became very significant to observe repeatedly what I had noticed on this first walk—how something can appear like a beautifully arranged garden.

Even someone who cannot achieve, or does not wish for, the kind of social dealings with nature-spirits touched on here, may observe the visible things here described and make surprising discoveries. And some contact with these beings is really coming about, when one concerns oneself directly with their activities.

When we had finished with the rock, I was led quickly to a place on the bank of the stream opposite a thickly grown island. The water was quite wide but I was directed to cross it. I resisted this—Steiner speaks of gnomes that play tricks! But I was persuaded and, with the help of a stepping-stone, crossed dry-shod. (Similar things often happened later.) On the island I was led on a devious path and shown several instructive things. Delightedly they pointed out a tall plant with brilliant Autumn colors, while everything else was still dull green. I asked myself—how can a gnome, who lives generally within solid things, be interested in my impression of the color of a plant? But this helped me especially to understand that the world is a scene enacted for us. As with a play, we should not hope to find the essential reality behind the scenes (as science too often does)—behind theatrical scenery there is generally ugliness. Just as the performance is the main thing with a play, to which everything else is subordinate, nature itself is a performance that is only successful if we look at it in the right way. If we understand this, the whole world becomes different for us.

During the entire excursion, I had looked ahead as little as possible, so as not to be distracted by questions of direction. And on the island it was such a tangle that I could not even see immediately ahead. At last I was led firmly toward the water and began to wonder about reaching the opposite bank. But just where I reached the water, there was a plank on which I could cross. This did not solve everything. The bank was a swamp—but I let myself be led by a zig-zag, without looking round anxiously for something better, to firm ground. Then there was a steep slope, to be negotiated by the ordinary public only on hands and feet. But I was brought like a guest of honor to a natural staircase, plain but secret, that I could ascend royally upright.

In the Whitsun lecture at Cologne (June 7, 1908), Rudolf Steiner describes how in the past gnomes used to help people with all sorts of things—mineworkers, for instance, in their search for metals. There and then too they pointed out the way. What a light is thrown by such possibilities on the powers of orientation shown by animals. They have no barrier of intellect between their will and the nature-spirits. Later on I often encountered much worse bogs, which I would never have been able to cross without help. If I let myself be led, I found the most unlikely ways through. Paths everywhere become visible when pointed out—but they never go in straight lines.

Again I was led to the bank of the stream—and there I had to stop for a long time. Several times already I had been held up, where there was nothing conspicuous to be seen. I understood this as a wish for more stories, and as soon as I complied, the expedition continued. But there was nothing fresh to be seen here, and stories had no effect. I accepted imprisonment and waited for my destiny. Above the ridge of the mountain on the other side the light was becoming more intense, and the white flaming edge of the sun emerged from the dark rock. This produced a mood of incredible festivity in the water rushing and leaping over the stones. This was what I had been led to see. Everywhere gleaming sparks, interweaving, mirrored, suddenly darkened. This was the climax of my expedition, which was, all of it, a festival. But how had all this been possible? It is a difficult job to be a guide—how could these beings be so good at it?

That day was just the beginning of my "gnome university." The indications were always of the kind I have described. The picture appeared magically in my mind, or outside—the rest I had to do myself. However simple the indications often were in themselves, they might point to far-reaching connections, even glimpses of the past and future of the earth.

Such an expedition brings about an extraordinary mood. One loses connection with time, and one's entire surroundings take on a fairy-tale quality. One is led into a strange kingdom. Yet on such paths I have never seen anything that could not be seen by anyone else—though perhaps things that nobody else has ever seen. Here the ideal of a science is unveiled, in harmony with the tasks and intentions of the nature-spirits. Such a science will no longer destroy fairy stories, but be a source of inspiration for them, for stories that are exact and true (as most of the old ones are), and can stand up to all the demands of a new, healed science.

Translated and abridged from a series of articles by Gisela and Adam Bittleston in the "Mededelingen" of the Anthroposophical Society in the Netherlands.

Appendix 3
A Selection of Birthday Verses
organized by theme

Earth

Tall and silent stands the mountain
Towering upward to the sky.
Bleak and barren is the summit
Where the lonely eagles fly.

Yet beneath the rocky hillside
Caverns hide their treasures bright.
May they someday be discovered,
May I bring them forth to light.

Within the stone lies hidden
A crystal clear and bright;
It waits in darkness for the day
When it will come to light.

Within my self lie hidden,
In depths of soul concealed,
The treasures that by Spirit-light
Will one day be revealed.

A diamond is fashioned through timeless ages
Deep in the heart of the earth.
So everything precious is wrought through great effort
And never brought quickly to birth.

Deep within the darkened earth there shines a secret light,
Cast by crystals in their caverns gleaming pure and bright.
So spirit-light illuminates its darkened earthly home,
And shines within the human soul like crystals in the stone.

The gold that has the greatest worth
And shines with purest light,
Cannot be found in depths of earth
Nor on the mountain height.

But those who look with loving eyes
Can learn how to behold
What lies within the human heart –
The precious, inner gold.

The hoard of gold that is kept in a cave
And hidden far from sight
Will be found one day and brought to light
And used for the good, the truth, and the right.

Water

Dark of water,
Light of air,
Weave the rainbow
Fine and fair.
So our souls
Weave patterns bright
From dark of earth
And Spirit-Light.

Softly the dew falls on stem, leaf, and flower,
Glist'ning like jewels in the dawn's early light.
Whence comes this delicate, radiant shower?
Why must it vanish so quickly from sight?

The pool that once was ruffled
By every little breeze
Now lies so placid, calm, and still,
Contented and at ease.
And in her deep dark waters,
Like diamonds clear and bright,
The countless stars of heaven gleam
Throughout the Summer night.

Slow and silent flows the river,
Ever onward to the sea.
Riches hide beneath the surface
Of its waters dark and deep.
Would that we might someday know
What lies hidden there below.

Swiftly and merrily rushes the stream down the mountainside.
Bubbling and splashing, its waters agleam, it turns the great mill wheel.
So may I too discover the way to work with all my might –
To seek the true, to do the good, and to uphold the right.

Above the green and fertile fields a cloud goes sailing by,
Frisky as a little lamb that frolics 'cross the sky;
But when the parched and thirsty plants implore the cloud for rain,
She lets her living waters flow and bathes the earth again.

Hard against the rocky shore
the swirling waters dash,
Wearing down the rugged stone
with every surge and splash.

Air

Lightly and merrily dances the breeze,
flecking the waves with white;
Blowing and billowing over the seas,
frolicking day and night.
But when their strength is needed,
the winds can lend their force
To drive ships swiftly o'er the waves
upon their chosen course.

Lightly the breeze dances over the meadow,
Cheerfully greeting the birds and the bees;
Bringing glad tidings to all of her friends –
The grains and the grasses, the flowers and trees.

Fire

Bright burns the fire, with fierce, raging flames,
Flames that can turn dross into gold.
Gold may be forged by heart's inner fire.
Fire of my heart – temper my will,
Fill me with light, fire burning bright.

The flames leap high,
Bright sparks reach the sky,
While softly below
The red embers glow.

Ember, flame, and spark –
Warm thou my thought,
Enlighten my heart,
That good may arise
From all I do.

In the cold and the darkness a fire is blazing,
High leap the flames, so daring and bright;
While gently below the embers are glowing,
Giving us warmth as well as light.

In the night a lantern's gleaming,
shining warm and bright –
Driving out the cold and darkness,
filling hearts with light.

Sun and Seasons

The light that permeates my thought,
The love that warms my heart,
The strength that courses through my limbs,
Have all been given me
That I may do whate'er I can
To help and serve my fellow man.

A Selection of Birthday Verses

The Light of the World came down to earth
To shine for all Mankind;
I carry that light within myself,
In my will, in my heart, and my mind.

O may my heart forever be as gentle as the dawn,
As strong and full of courage as the blazing sun at noon;
As patient as the setting sun, so radiant, warm and bright;
As tender as the twilight that welcomes in the night.

In the cool of the morn
The mist lingers on,
Wreathing the hills with white;
Until the sun,
So gentle and warm,
Fills all the earth
With his glorious light.

High above the fertile earth
The sun is shining bright,
Quick'ning every living thing
With warmth and golden light.
Deep within the human heart
There shines another sun,
Kindling love and living thought
So that His will be done.

Throughout the cold, hard Winter,
The icicles hang down,
Bedecking every snowy roof
With a glist'ning, crystal crown.
But when the days grow warmer,
And geese are on the wing,
The icicles melt till they fall with a crash
That welcomes in the Spring.

Too long the cold, hard Winter
Has gripped the frozen earth;
But now the sun at last has come
To herald Spring's rebirth.

O sun who warms the earth again
Come now and warm my heart!
Awaken me, abide with me,
And nevermore depart.

Moon and Stars

The beautiful, radiant evening star
has such a lovely voice,
That whenever she joins the heavenly choir
the angels all rejoice.

The crescent moon, so meek and mild prefers to hide her face,
But when she waxes right and full, she shines with strength and grace.
O may I too develop strength to shine with all my might –
That like the lovely silver moon, I fill the dark with light.

Through the long Winter nights, though darkness held sway,
The stars in their courses shone steady and bright.
So too in my life, when dark be the way,
I know I can count on my soul's inner light.

Plants

Awake! Awake! The sun calls out
To rouse the slumb'ring earth.
Then all the Springtime flowers bloom
In glorious rebirth.

From little seed the sunflower grows tall and straight and strong;
With cheerful face she greets the sun all through the Summer long.
When Autumn comes her petals fade, her leaves grow brown and old;
But birds will feed upon her seeds throughout the Winter cold.

A Selection of Birthday Verses

In the ground so dark and deep,
Seeds lie in their Winter sleep,
Safely in their beds of earth,
They await the Spring's rebirth.

Spirits of the earth, awake!
Pray do not thy task forsake.
Bring an end to Winter's night,
Drive the darkness out with light.

Within the seed lies hidden
The plant that one day will become –
The sturdy roots, the graceful leaves
The flowers reaching toward the sun.

Within the soul lies hidden
The seed of all that I could be.
O sun, who calls the flowers forth –
Call forth the higher self in me.

Like the lily pure and good, like the moon so bright –
May my every thought be filled with beauty, truth, and light.
Like the rose so rich and red, like the sun above –
May my every word and deed be filled with strength and love.

Lovely the flowers that grow in the meadow,
Fragrant and sweet 'neath the warm Summer sun;
Precious the grains and the grasses beside them,
Yielding their bounty when harvest is come.

Like a flower whose petals close tight for the night
But who opens them up to the sun's loving light,
May my heart open up to the sun every day,
That by its light I find my way.

Trees

Though blossoms will fade and blow away
And leaves are shed every Fall,
Though branches may crack and break someday,
The tree stands straight and tall.

Like the roots that grip the ground
And hold the tree upright –
May my will be strong enough
That I may continue to reach for the Light.

Over the stream the willow
Bends down with tresses long,
And listens pensively all day
To the water's lovely song.
The sunlight-dappled water,
The moonlit silver stream,
Reveal their mysteries to her
Like echoes of a dream.

The slender little sapling
Has roots so strong and sure;
Though storm winds blow and bend him low,
He knows he will endure.
And all that he must suffer
And bear throughout the year
Will help him be a mighty tree
Who knows not doubt nor fear.

A sudden storm tears through the trees,
hurling branches down;
But one tree braves the wind with ease
by bending to the ground.
When at last the storm is past
and gone the wind and rain,
The tree that weathered every blast
stands upright once again.

A Selection of Birthday Verses

Tall grows the oak, with roots deep and strong;
Straight grows its trunk, with limbs spreading wide.
O may my heart grow strong like the oak;
O may my limbs grow strong to do good.

Majestic stands the mighty oak
With broad and leafy crown;
Yet Winter winds will strip the boughs
And strew the leaves around.
But sturdy roots support the trunk
And hold it sure and fast,
So that the oak can stand the storms
And weather Winter's blast.

Green was the tree all Summer,
But now she's dressed in gold,
And stands majestic as a queen
Who reigned in days of old.

So may the gold within my soul
Be brought to light one day,
And all my inner treasures
Be well and freely shared.

I brave the cold of Winter
Endure the Summer's heat,
My leaves are strewn by Autumn storms,
My branches bent by sleet.
Yet every Spring my flowers bloom,
My fruit grows ripe each Fall;
Despite all this adversity
I still stand straight and tall.

In the forest deep and dark,
Trees stand silent, bare, and stark;
Yet there's one whose branches gleam
In their cloaks of living green.
O may I too shine through the year,
And like the spruce bring joy and cheer.

Animals

Majestic is the tiger,
So noble, brave, and bold;
He roams serenely through his realm
In robes of burnished gold.
And though he wanders all alone
Through the somber forest glade,
He does not mind the solitude
And never feels afraid.

The timid, little woodland fawn
grows stronger year by year;
A mighty stag he will become,
who knows not doubt nor fear.

O may I too grow through the years
in courage, strength, and grace
That, like the mighty woodland stag,
I find my rightful place.

Strong and swift the dolphin,
Who swims so gracefully;
Far from land she ventures forth
Across the trackless sea.
Down she dives beneath the waves
To explore the depths alone;
Then up to the light she dashes again
To frolic in the foam.

Birds

How sweetly sings the nightingale all through the Summer long;
Delighting everyone who hearkens to his lovely song.

In the dark before dawn a lone voice resounds
Breaking the stillness of night.
First one, then the other birds answer her call,
In jubilant chorus to welcome the light.

A Selection of Birthday Verses

Swiftly and surely the barn swallows fly,
Winging their way through the warm Summer sky.
Like a graceful barn swallow may I ever be,
Bringing joy and delight to all who know me.

How gracefully the great geese fly on strong and sturdy wing;
Heading southward in the Fall, returning in the Spring.
Straight and swift as arrows, they speed upon their way,
Always knowing where to go by night as well as day.

Like a fledgling who is just beginning to fly,
And who sees all the other birds soaring so high –
I know the time will not be long
Till my wings too grow sure and strong.

The little cygnet on the lake is awkward, meek, and shy;
For, till she is completely fledged, she cannot learn to fly.
But when she grows into a swan, she glides so gracefully,
And soars aloft on mighty wings of regal majesty.

High above the mountain peaks
The eagle soars alone;
Borne aloft on might wings,
Circling heaven's dome.
Naught escapes his piercing gaze,
None his grasp defies,
Bold he flies so free and fast –
Master of the skies.

Loud the wind and hard the rain
fiercely blows the storm;
Yet the dove within her nest
is cozy, safe, and warm.

And though the night is cold and dark
and drenched with bitter rain,
She sleeps serenely, for she knows
the sun will shine again.

Insects

The honeybee flies through the orchards and meadows
In search of sweet nectar from flowers and trees.
When she returns to the hive with her treasure,
She shares the rich honey with all of the bees.

Like a beautiful, delicate butterfly,
Who in search of sweet flowers goes fluttering by –
I will search for life's treasures where e'er I alight;
May I cherish and guard them and use them aright.

The caterpillar long awaits
Her moment of rebirth,
When she can spread her radiant wings
And soar above the earth
To seek the sun who called her,
And raised her from her sleep,
When in her crystal casket
She lay in slumber deep.

Journeys

To walk along the level road requires little will,
But climbing up the mountainside develops strength and skill.
And those who choose to stay below may seek but never find
The riches that will be revealed to those who dare to climb.

The boulders and stones that lay strewn in the fields
Were once a burden to all.
To effort and patience they slowly did yield
And were made into sturdy stone walls.

So too may every obstacle
Be finally overcome,
Transformed into life's building stones
For all that must be done.

The long road is stretching before me,
Its distances beckon to me;
With courage and strength I stride forth at dawn
To seek my destiny.

The long road is stretching behind me
Its obstacles overcome;
But here from the heights I can clearly see
That my journey has barely begun.

Hard is the path that leads to the peak –
Rock-strewn and rugged, sharp, steep, and high.
Lofty the heights that hold what I seek;
Yet I will climb and reach for the sky.

Hard was the path
That led to the peak;
Yet I did strive
And reached for the sky.
Now I can see
That the treasures I seek
Will be forged in my soul
By each mountain I climb.

In days of old the wanderers' ways
Were lit by stars which shone so bright.
O May the paths I tread in life
Be lit by my own inner light.

Beyond the shore, the trackless sea
Is stretching far and wide;
Beyond the hills, the lofty peaks
Are rising to the sky;
Beyond each goal that I pursue
Another beckons me;
And I will strive forevermore
That farther goal to seek.

Work & Occupations
especially suitable for 3rd grade

Strong and steadfast is the plow horse,
Working patiently all day;
Forth he strides to meet each challenge,
Never falt'ring on the way.

My hands have been busy
My heart now is full;
The garden I planted
Has flourished and grown.

So too in the harvest
Of life may I reap
The well-ripened fruits
Of my thoughts, words, and deeds.

The evening sun is golden,
The sun sinks in the west;
The cattle calmly chew their cud,
The sheep lie down to rest;
Then sounding through the twilight,
A tune so soft and sweet,
Of a shepherd playing on his pipe
As he watches o'er his sheep.
And now the last notes linger,
Resounding o'er the hill;
The sky grows dark, the stars come out,
Then all, at last, is still.

The ground was well-tilled
And the seeds caref'lly sown;
The garden I planted
Has flourished and grown.
So too may my worthiest
Strivings take root
And ever may my
Good intentions bear fruit.

Like the farmer who works for the sake of his farm,
For the animals, plants, and the living land,
May I work for the sake of my fellow man,
Through the warmth of my heart and the strength of my hand.

Each day the gard'ner tends her plot
with strong and gentle hand;
And cares for every tree and plant
that grows upon the land.

So will I tend to my life's work,
and ever will I strive
To care for everything I do
that it may grow and thrive.

Through the year a man must toil
To reap the bounty of the soil.
Till the earth in sun and rain,
Harvest, bind, and thresh the grain.
Loving heart and steady hand
Reap the bounty of the land.
So may I ever do my part
Through the strength of my hand
And the warmth of my heart.

Carefully the nest is crafted,
Every stalk and stick in place,
For it must protect a treasure,
Keep it cradled, warm and safe.

The carpenter works carefully
On every piece of wood;
Each cut he makes is straight and true,
Each joint is strong and good.
And though he is meticulous,
He keeps his goal in mind,
So that his work is always done
By the appointed time.

The mason shapes and lays the stones
With sure and steady hand.
The walls he builds with strength and skill
For centuries will stand.
So may I lay within my soul
In thought and word so true,
A lasting, firm foundation
For all that I must do.

Appendix 4
Suggested Resources

Nature Stories and Fables

Aesop, Illustrated by Milo Winter. *Aesop's Fables for Children*. Chicago: Rand McNally, 1919. (Reprinted by various publishers. Project Gutenberg/Open Library).

Aesop, Olivia & Robert Temple, translators. *Aesop – The Complete Fables*. London: Penguin Books, 1998.

Andrews, Jane. *The Stories Mother Nature Told Her Children*, CreateSpace, 2013.

Cook, Flora J. *Nature Myths and Stories for Little Children*, CreateSpace, 2014.

Copple, Rudolf. *To Grow and Become: Stories for Children*, Hudson, NY: Waldorf Publications, 1994.

Gmeyner, Elizabeth and Joyce Russell. *The Key of the Kingdom: A Book of Stories and Poems for Children*. Great Barrington, MA: Bell Pond Books, 2004.

Harrer, Dorothy. *Nature Ways in Story and Verse*. Spring Valley, NY: Mercury Press, 1986.

_____. *Verses and Poems & Stories to Tell*, Hudson, NY: Waldorf Publications, 2014.

Holbrook, Florence. *The Book of Nature Myths for Children*, Kindle, 2022.

La Fontaine, Jean. *Fables*. Project Gutenberg/Open Library.

Patten, William. *The Junior Classics, Volume 2: Folk Tales and Myths*, CreateSpace, 2017; originally published New York: P.F. Collier & Son Corp., 1918.

Peckham, Margaret. *Nature Stories*. Spring Valley, NY: Mercury Press, 1982.

Pierson, Clara Dillingham. *Among the Forest People, Among the Meadow People, Among the Night People,* and *Among the Pond People*, CreateSpace, 2014; originally published Boston: Houghton, Mifflin and Co., 1902.

Schwartz, Eugene. *Why the Setting Sun Turns Red and Other Stories*. Fair Oaks, CA: AWSNA Publications, 2004.

Skinner, Ada M. & Eleanor. *The Emerald Story Book: Stories and Legends of Spring, Nature and Easter*, Kindle, 2009; also London: Forgotten Books, 2017.

_____. *Little Folk's Christmas Stories and Plays*, Hardpress Publishing, 2013.

_____. *The Pearl Story Book: Stories and Legends of Winter, Christmas and New Year's Day*, Ithaca, NY: Yesterday's Classics, 2021.

Streit, Jakob. *What Animals Say to Each Other: 30 Nature Fables in Rhyme*. Chatham, NY: Waldorf Publications, 2014.

Tolstoy, Leo. *Fables for Children, Stories for Children, Natural Science Stories, and Moral Tales*. Project Gutenberg/Open Library.

Wiggin, Kate Douglas and Nora Archibald Smith. *Talking Beasts – A Book of Fable Wisdom*. Project Gutenberg/Open Library.

Multicultural Tales and Legends

Bierhorst, John. *Latin American Folktales – Stories from Hispanic and Indian Traditions*. New York: Pantheon, 2003. Also available on Open Library.

Clark, Kate McCosh. *Maori Tales and Legends. Collected and Retold*. City, Franklin Classics: 2018.

Cowell, Edward Byles. *The Complete Jataka Tales*. North Charleston, SC: Createspace, 2016.

Dharma, Krishna. *Panchatantra – A Vivid Retelling of India's Most Famous Collection of Fables*. Kindle edition from Amazon; also available paperback, Torchlight Publishing, 2004.

Ferguson, Gary. *Spirits of the Wild – The World's Greatest Nature Myths*. New York: Crown Publishers, 1996.

Honeij, James. *South African Folk Tales*. Project Gutenberg/Open Library.

Kanawa, Kiri Te. *Land of the Long White Cloud: Maori Myths, Tales, and Legends*. New York: Arcade Publications: 1990.

Poems

Burton, Michael Hedley. *In the Light of a Child*, Ghent, NY: Adonis Press, 1989.

Kennedy, David. *The Waldorf Book of Animal Poetry*. Living Arts Books: Viroqua, WI, 2013.

_____. *The Waldorf Book of Poetry*. Living Arts Books: Viroqua, WI, 2012.

Skinner, Ada M. & Eleanor. *A Child's Own Book of Verse, Book One*, Ithaca, NY: Yesterday's Classics, 2006.

Thomas, Heather. *A Journey Through Time in Verse and Rhyme*. UK: Floris Books, 1998.

Music

Adams, David. *One, Two, Three: A Collection of Songs, Verses, Riddles and Stories for Children of Grades 1-3*, Hudson, NY: Waldorf Publications, 2003.

Foster, Nancy. *Let Us Form a Ring*, WECAN Publications (Waldorf Early Childhood Association of North America), Acorn Hill Children's Center, 1998.

Hall, Doreen. *Music for Children*, New York: Associated Music Publishers, Schott Music Corp., 1960.

Landeck, Beatrice (ed). *More Songs to Grow On: A New Collection of Folk Songs for Children*, Florence White (arr.), New York: Edward B. Marks Music Corporation, 1954.

Lebret, Elisabeth. *Pentatonic Songs for Nursery, Kindergarten and Grades 1&2*, Waldorf Association of Ontario, 1985.

Logan, Arnold. *A Garden of Songs for Singing and Piping at Home and School*, Hudson, NY: Waldorf Publications, 1996.

Resnick, Jeanette. *Voyage in Song and Story: Music, Poetry and Plays for Teachers and Children Kindergarten–Grade 6*, Hudson, NY: Waldorf Publications, 2018.

Riley, Alice D. and Jessie L. Gaynor. *Songs of the Child World*, Cincinnati: The John Church Company, 1927.

Schweizer, Sally. *Spring Songbook: Seasonal Verses, Poems and Songs for Children, Parents and Teachers: An Anthology for Family, School, Festivals and Fun!* Other anthologies available for Summer, Autumn and Winter. Forest Row, UK: Rudolf Steiner Press, 2018.

Seeger, Ruth Crawford. *American Folk Songs for Children in Home, School and Nursery School*, Garden City, NY: Doubleday, 1948.

Swinger, Marlys. *Sing through the Day: Eighty Songs for Children*, Walden, NY: Plough Publishing House, 1999.

_____. *Sing through the Seasons: Ninety-Nine Songs for Children*, Walden, NY: Plough Publishing House, 1986.

Winn, Marie (ed). *Fireside Book of Children's Songs*, Allan Miller (arr.), New York: Simon & Schuster, 1966.

Yolen, Jane (ed.) *The Fireside Song Book of Birds and Beasts*, Barbara Greer (arr.), New York: Simon & Schuster, 1972.

Crafts

Berger, Petra. *Feltcraft*, Edinburgh, UK: Floris Books, 2010.

Berger, Thomas. *Christmas Craft Book*, Edinburgh, UK: Floris Books, 2011.

_____. *Harvest Craft Book*, Edinburgh, UK: Floris Books, 2011.

Berger, Thomas & Petra. *Crafts through the Year Book*. Edinburgh, UK: Floris Books, 2011.

_____. *Easter Craft Book*, Edinburgh, UK: Floris Books, 1993 (out of print).

_____. *Gnome Craft Book*, Edinburgh, UK: Floris Books, 2010.

Bowden, Marcia. *Nature for the Very Young – A Handbook of Indoor and Outdoor Activities*. Hoboken, NJ: Wiley, 1989.

Seasonal/Food

Carey, Diana and Judy Large. *Festivals, Family and Food: A Guide to Seasonal Celebration*, Stroud, UK: Hawthorn Press, 1982.

Druitt, Ann, Christine Fynes-Clinton and Marije Rowling. *All Year Round*, Stroud, UK: Hawthorn Press, 1995.

Hildreth, Lisa. *The Waldorf Kindergarten Snack Book*, Great Barrington, MA: Bell Pond Books, 2005.

Huff, Andrea. *The Waldorf School Book of Soups*, Great Barrington, MA: Bell Pond Books, 2006.

Mihyo, Paschal. *The Birthday Book: Celebrations for Everyone*, Stroud, UK: Hawthorn Press, 2004.

Post, Marsha (ed). *The Waldorf Book of Breads*, Hudson, NY: SteinerBooks, 2009.

Rosengren, Rikke and Nana Lyzet. *Pumpkin Soup and Cherry Bread: A Steiner-Waldorf Kindergarten Cookbook*, Edinburgh, UK: Floris Books, 2015.

Made in the USA
Middletown, DE
22 October 2022